Good Cat!

A Proven Guide to Successful

Litter Box Use

and Problem Solving

Shirlee Kalstone

Illustrations by John Martin

Howell Book House
Published by Wiley Publishing, Inc., Hoboken, New Jersey

Library of Congress Cataloging-in-Publication Data:
Kalstone, Shirlee.
  Good cat!: a proven guide to successful litter box use and problem solving/Shirlee Kalstone;
illustrations by John Martin.—1st ed.
      p. cm.
  Includes index.
  ISBN 0-7645-6936-8 (pbk.: alk. paper)
1. Cat litter boxes. 2. Cats—training. 3. Cats—Behavior. I. Title.
  SF447.34.K35   2004
  636.8'0835—dc22
                                                          2004019708

Printed in the United States of America
10 9 8 7 6 5 4 3 2 1

Book design by LeAndra Hosier
Cover design by Wendy Mount
Book production by Wiley Publishing, Inc. Composition Services

Contents

Acknowledgments

I would like to thank the following organizations, journals, magazines and individuals: the American Society for the Prevention of Cruelty to Animals (ASPCA); Humane Society of the United States (HSUS); Centers for Disease Control and Prevention (CDC); Association of Specialists in Cleaning and Restoring (ASCR); American Pet Products Manufacturing Association (APPMA); *Cat Fancy*, *Pet Age* and *Pets International* magazines; *Catnip* (Tufts University School of Veterinary Medicine); *CatWatch* (Cornell University College of Veterinary Medicine); Cornell University Feline Health Center; Morris Animal Foundation; Farnam Companies; Bonnie V. Beaver, D.V.M., M.S.; Wayne Hunthausen, D.V.M.; Bruce Fogle, D.V.M., M.R.C.V.S.; Dr. Nicholas Dodman, professor of Behavioral Pharmacology, Tufts University School of Veterinary Medicine, and director of the Behavior Clinic; Paul H. Schwartz, D.V.M. (and the Center for Veterinary Care in Manhattan for taking such great care of our cats); and Don Aslett.

Very special thanks to my agent, Joan Raines, for encouraging me to start writing again. Above all, thanks to my husband, Larry, for his constant support. He's always been there for me, most especially during a recent serious illness.

Introduction

In the past decade, cat ownership has increased dramatically while dog ownership has declined. The latest American Pet Products Manufacturers Association survey shows 77.7 million cats living in American households. More and more cats are also living indoors, and today's indoor cats are living longer. According to the Humane Society of the United States, an average outdoor cat's life expectancy is between 2 and 4 years, while that of an indoor cat is 17 years or more. Cat ownership has skyrocketed not only in the United States but also around the world. Euro-Monitor International, publishers of market research reports, estimates the global cat population is now slightly more than 200 million.

Cats make ideal pets for several reasons: they are clean and fastidious creatures, they are affectionate, they are flexible in their care needs, they don't have to be walked (cats can live long and happy lives without ever going outside), they are much quieter and less aggressive than dogs, they cost less to maintain than many other kinds of pets and they are easily trained to use a litter box.

Even so, many owners seem unable to cope with their cats' house-soiling habits. *More cats end up in shelters each year because of inappropriate elimination problems than from any other cause.* Although certain medical and physiological conditions can cause litter box avoidance and inappropriate elimination, most litter box problems crop up when owners fail to understand their cats' needs.

Litter box training usually is very easy because cats have a natural tendency to bury their waste. *Cats do not normally urinate and defecate outside their litter boxes unless something is wrong.*

When elimination problems happen and medical causes are ruled out, it becomes obvious that the cat is trying to tell her owner that something is wrong. There are many reasons cats avoid the litter box and urinate or defecate outside it: a dirty box, a dislike for a certain type of litter, inappropriately positioned box, too few boxes, animosity between cats in the house, anxiety or stress—and there are many others.

To solve these problems, you must figure out what your cat is trying to tell you. As one expert writes on the Internet, "Cats use elimination as a communication—a kind of pee-mail, if you will." Exactly! Your cat is not being vindictive. She is communicating with you using the only language she knows. It's your job to translate, and this book can help.

Punishing the cat for her behavior (especially in the vicinity of her litter box) does not work and will not solve the problem. Actually, it could make things worse because a negative experience associated with the litter box could inspire the cat to avoid the box entirely.

In the wild, your cat keeps her territory clean by eliminating far from the places where she spends her time, and by burying her waste. But in your home, your cat must rely on you to give her a clean, safe place to do her business. *Good Cat!* tells you everything you need to know about litter boxes: the different types of boxes and accessories you can buy, the best size and type of box for your cat and where it should be located, the pros and cons of the many different kinds of litter, how often the litter should be changed, how to control odor, tips to prevent litter box problems before they start and the best methods for cleaning and deodorizing urine stains. More important, this book explains all the reasons—medical, territorial, behavioral, emotional and environmental—for inappropriate elimination and gives you advice about how to solve these problems.

Understanding Feline Behavior

Cats, to the casual observer, are often thought to be independent, aloof creatures who spend most of their time in solitary pursuits. But the truth is that cats are relatively social animals who enjoy our company and use a host of communication skills to express their desires and intentions to one another and to us.

However, cats do seem to have been less affected by domestication than dogs. They can be loving and devoted creatures who delight in the comforts of affection, regular meals and an agreeable home, yet they still retain the inborn hunting instincts of their early ancestors. Remember, cats were originally domesticated to hunt and control vermin—and to do it on their own, with no direction from humans. Dogs, on the other hand, have been selectively bred for millennia to work closely with humans at a variety of jobs: hunting, herding, guarding, as dogs of war and as companions. Although cats have also been associated with humans for thousands of years, they rarely have been intentionally bred for specific purposes other than vermin control. Even though they come in a variety of colors and coat patterns, their conformation and natural hunting tendencies have remained basically unchanged since ancient times.

Dogs and cats are different in another important way, too. Dogs are social animals who evolved to live in packs and are therefore genetically hardwired to conform to pack behavioral patterns. Pack relationships are based on a hierarchy where there is a pack leader who always dominates, disciplines the rest and maintains group order. The same instinct for pack behavior governs the dog's relationships with humans. People become pack members, just as if they were dogs. Dogs are highly motivated by and dependent on their leaders or masters, and this makes them very trainable. When they are disciplined or punished by their masters, they exhibit submissive behavior—just as they would to the leader of the canine pack.

The Feline Hierarchy

Cats, on the other hand, do not form pack hierarchies, where one animal is always clearly dominant. Aside from the interplay between a mother and her kittens, or brief encounters during mating season, much time is spent avoiding one another. The rituals of scent marking, described later in this chapter, also help to reduce close contact between cats. In fact, all members of the wild cat family, with the exception of lions (who live in social groups known as prides), can be considered loners. While there has been some research to show that feral cats willingly form colonies, they are not close-knit groups the way dog packs are. Domestic cats, therefore, do not respond to people as if they were pack leaders or members. It is futile to try to dominate or even train a cat as one would a dog. Rather than assuming a submissive posture in response to discipline, the cat will *always* try to escape or fight.

This does not mean, however, that cats are totally asocial. Although there are no *fixed* hierarchies of dominant cat, second cat and so on down the line, cats often form what animal behaviorists call *relative* dominance hierarchies that are related to time and place. For example, during play or play-fighting, veterinarian Bruce Fogle writes in *The Cat's Mind*, kittens may take turns being the dominant one. Or one cat may be domineering at mealtime, while another rules over the litter box.

Cats can develop enduring social relationships with other animals as well as people. Cats who live from kittenhood with other cats or dogs can become very affectionate and protective toward these companions, and the longer the animals live together, the stronger the relationship becomes. Cats can also mourn the loss of a feline, a canine or a human companion, and these deep feelings can affect their litter box habits.

Territory and Aggression

Domestic cats, like feral and wild cats, are territorial creatures. In the wild, cats establish their territories based on the number of cats and the amount of food and shelter in a given area. They will defend their territories, when necessary, to keep other cats from killing their prey, to protect their young, during the mating season and simply to keep out intruders.

Domestic cats do not always choose their territories; generally, the territory is chosen by a cat's owner instead. Each cat, however, will have his home base. This is usually a room or a favorite corner of a room in the house where the cat lives. Around the home base (the rest of the house, or the house and yard, if the cat is permitted outdoors) are areas the cat likes to use for napping, playing, sunbathing and surveillance. The extent of the home base depends on the age and temperament of the cat, and especially the sex of the cat and whether the cat is neutered. Females and neutered cats of both sexes seem to feel more content within a limited area of their home or yard, which they will spiritedly protect. The home range of unneutered males may be many times larger, especially during the mating season.

If a cat is permitted to roam, beyond his limited outdoor home base is a range connected by an elaborate network of pathways leading to more or less regularly visited areas for hunting, courting, contests and fighting and other activities.

In most cases, the boundaries of both outdoor and indoor territories are firmly established. Within them, a stranger must be prepared to challenge the resident cat; outside the boundaries, the intruder will be overlooked. Many cats are satisfied to spend their entire lives indoors, but these territorial imperatives still apply. An indoor cat will still assert ownership and defend a favorite location, for example, a part of the house, a piece of furniture or a window seat. And his bailiwick may also include the yard, if it is observable from the window.

Indoors, a cat's aggressiveness in defending his territory may not be so obvious in a single-cat home. But when there are two or more cats, the territorial imperative becomes more clear as each cat determines his home base and learns to share other areas with the other cats.

A confrontation indoors can be serious when one cat invades the territory of another. When his territory is threatened, a dominant cat will usually try to intimidate the intruder using aggressive behavior: hissing, growling and screaming vocal threats, baring his teeth and assuming offensive body positions (discussed at length later in this chapter). If these postures and physical intimidation do not scare off the intruder, the dominant cat may resort to other destructive practices, including urine marking and/or urinating and defecating outside the litter box, to reconfirm precisely who is the top cat. Conversely, if the litter box happens to be in an area that one cat in the household claims as his own, the other cats may be too intimidated to use it. Solutions for these behaviors are offered in chapter 5.

How Cats Communicate

Cats have an extensive vocabulary. They communicate with body language and voice, with visual and olfactory marks. They use their face, eyes, ears, whiskers, body, paws, tail, fur, posture, voice, urine and feces to express their feelings and to deal with other cats, other pets and humans. The complex combinations of the body postures they adopt, the sounds they make and the places where they rub, scratch and eliminate all play a role in expressing how they feel and what they want. They can show in *very* specific ways when they are happy and contented or angry or stressed. Understanding what your cat is trying to tell you is important in strengthening the bond between the two of you. It also is important in solving litter box problems—because you can't fix the problem until you know what it is.

Body Language

Cats have a variety of facial expressions in which the eyes, ears and whiskers play an important role. When a cat is happy and content, she will sit with her face relaxed, ears upright and eyes partly closed or with the pupils narrowed to a slit. A cat who is being stroked and spoken to will keep her eyes this way while purring and turning up the corners of her mouth in a sort of smile. The pupils of an angry cat, or one facing an opponent, will dilate, the ears flatten to the sides of the head and the mouth opens to express a warning. Any intense emotional stimulus, such as anger, fear, pleasure, agitation or excitement, can cause the pupils to contract suddenly.

The position of a cat's ears is another mood indicator. Erect ears that face forward express relaxation. A curious cat will prick up her ears and push them slightly forward to focus on sound. Ears held back with the body held low to the ground signal caution or reluctance. When a cat feels threatened, the ears turn to the side. Be wary, however, when the ears go down; if a cat is really angry or terrified, the ears are completely flattened against the head to shield them from an opponent's teeth and claws should a fight ensue.

The cat's whiskers are long, stiff hairs (otherwise known as *vibrissae*) embedded in extremely sensitive follicles in the skin above the eyes, on the cheeks and upper lips and on the backs of the forelegs. They function primarily as sensory devices—antennae, more or less—helping a cat detect the presence, size and shape of objects and obstacles close up, in restricted spaces and in the dark. The whiskers also play a role in communication with people and other cats. Fanned out whiskers indicate the cat is confident, relaxed and probably approachable. Whiskers that are fanned forward indicate curiosity. When a cat is agitated, frightened or ready for a fight, the whiskers are pulled backward and flattened against the face.

The tail is also another way cats communicate their moods. As a rule, the higher the tail, the better the cat's mood. A tail held very straight and high can be a form of greeting or a sign of pleasure. A cat who holds her tail erect can also be saying "I'm hungry" as she looks forward to a meal. A tail arched over the back or into an inverted U means the cat is merry and playful, but a tail arched downward means aggression. Some cats swish their tails from side to side when you talk to them or when they are pleased, but lashing or beating the tail back and forth from its base indicates tension or anger. The more rapid the swish, in fact, the more upset the cat. A tail carried low or tucked between a cat's legs is a sign of fear or submission.

Cats use various body postures to tell other cats or individuals whether they welcome a closer approach. Rolling over and expos-ing the abdomen or tilting the head is felinese for "I want to play." During the breeding season it can also mean (in an unneutered female) "I want to mate." Contentment or relaxation is expressed by several positions, including lying stretched out on one side and sitting with the paws deftly folded underneath and the tail curled around the body. The classic "Halloween cat" silhouette with the cat turned sideways, back arched, tail stiff and puffed up and claws unsheathed is an extreme threat posture. The posture is further enhanced by the facial expression: dilated pupils, whiskers held close to the face, ears flattened back, lips drawn back and teeth bared. The idea here is for the cat to look big and fierce, as if to say, "I don't want to fight, but I will if you come too close."

An offensive threat posture indicates that a cat is fearless and likely to attack. She faces her assailant head-on, in a straightforward stance, making direct eye contact and attempting to stare down her adversary. The whiskers fan straight out and the ears are flattened back. The tail lashes from side to side. When two cats are in this kind of standoff (and they can maintain this posture for 15 minutes or longer), they hiss and scream at each other until ritualized fight-ing begins, or until one becomes intimidated and capitulates. Cats capitulate by flattening their bodies on the ground, legs and feet tucked beneath them, with ears flattened and tail pulled in tight to show submission. If an intimidated cat can spot a way to escape, she will make a quick exit. If the cat is backed into a corner, however, and can no longer run away, she most likely will assume the defensive "Halloween cat" posture or a crouching posture. Don't mistake a cat rolling onto her side and extending her rear legs as a sign of submission; this is a posture that says the cat is ready to fight—the rear legs are extremely effective weapons.

A sick or desolate cat has a woeful facial expression. She carries her tail low and hunches up her body. She may not eat or clean herself, may vocalize more and will certainly be less playful.

Vocal Communication

Cats also use many sounds to express themselves vocally. Scientists have identified about 100 distinctive cat vocalizations. These vocal sounds are grouped into three patterns—murmurs, vowels and strained or high-intensity sounds—based on how they are produced.

The murmur patterns are sounds a cat makes while her mouth is closed; they include purring and the dulcet, trilling vocalizations that express greetings or acknowledgment. There are many theories as to why cats purr, and the mechanics of how they actually make the sound is still a mystery. Cats are able to purr in monotone in response to their mother licking them when they are two days old (this probably communicates to their mother that they are content and well fed). As kittens mature, the purring begins to vary in speed, pitch, rhythm and volume, producing many types of sounds. Generally, purring is a sign of pleasure and contentment, although some cats who are ill, severely injured, in pain, frightened or giving birth often purr resonantly, perhaps as a way of asking for help.

The vowel patterns cats make include different sounds, such as "meow" and its many variations, used by a cat to coax, demand, complain, inform and express surprise. Most of the chatty sounds made by Siamese are classified as vowel patterns. The sounds are started while the cat's mouth is open and finished when it is closed, and are used to communicate with other cats and with humans. Most cats develop a vocabulary of specific sounds that mean "please," "no," "food," "dirty litter box," "out," "play" and many others that you can learn to understand if you listen closely. Generally, the more aggravated a cat becomes, the lower the pitch of the meow.

Strained or high-intensity sounds are made with the mouth open and express anger or emotion. Used mostly for communicating with other cats, these include growling, snarling, hissing, spitting, yowling, screaming and the ritualistic mating cry. Cats make these sounds when they are frightened, angry, mating, fighting or in pain.

Each cat has her own particular vocabulary, the size of which will vary greatly depending on breed, sex and temperament. Siamese, Abyssinians and Oriental Shorthairs, for example, are known to be very talkative, while Persians tend to say very little. Cats carry on conversations with their owners, their kittens and other cats. But undoubtedly, vocal communication reaches its pinnacle during the mating season. Many unneutered females become very noisy when they go into heat and call loudly and constantly to inform the opposite sex that they are ready to mate, while the males howl and caterwaul at night.

Then there is the enchanting silent meow, where the cat opens her mouth and appears to meow, but no sound comes out. (The term was made popular in 1964 by Paul Gallico in *The Silent Meow*, a handbook written exclusively for cats, advising them on how to overpower a human family and discipline the family members.) Adult cats also do this. Although there are several theories as to the purpose of the silent meow, its meaning is known *only* to the cat who makes it.

An Illustrated Guide to Feline Body Language

To sum up this section, cats express their feelings, frame of mind, preferences and desires through their facial expressions, body postures and language. To better understand what your cat is saying, however, you should consider these *as a whole*. In other words, it is not useful to observe the position of the ears or the whiskers but ignore the arch of the back, the stance of the body, the angle of the tail, whether the pupils are dilated and if the fur is standing on end. The following illustrations show the most common postures and what they mean.

Alert, Happy, Content

- Face relaxed
- Ears erect, facing slightly forward
- Normal pupils, eyes wide open or partly closed
- Whiskers fanned out or slightly forward
- Mouth closed
- Tail held straight and high, fur flat and not bristling
- Standing erect, lying stretched out to the side or sitting with paws folded underneath and tail curled around body
- Mellow meow, purr or murmur

An alert, happy, content cat's expression and body posture

Fearful
- Dilated, very large pupils
- Whiskers pulled backward and flattened against the face
- Ears flattened sideways and backward
- Back arched
- Fur often stands straight out down back and tail
- Tail lashes back and forth or is held between legs
- Growling, hissing or spitting

A fearful cat's expression
and body posture

Defensive Threat
- Dilated, very large pupils
- Ears flattened back against the head
- Whiskers pulled back and flattened against the face
- Lips slightly open
- "Halloween cat" silhouette: body turned sideways toward adversary
- Fur on body and tail stands straight out to make the cat appear larger and more menacing
- Tail held straight up or close to the ground
- Spitting, hissing
- If aggressor comes near, the cat may assume a crouching position.
- In extreme cases, if the cat cannot escape, she will roll on her back and raise her paws in defense.

The Halloween cat posture

A defensive cat may crouch down

Extreme defensive posture, with claws at the ready

The expression and body posture
of a cat on the offensive

Offensive Threat

- Dilated, very large pupils
- Whiskers fanned straight out
- Ears flattened back against the head
- Lips drawn back into a snarl
- Fur on shoulders and tail bristles straight out to appear more menacing
- Direct eye contact with adversary
- Faces assailant head-on with hindquarters raised
- Tail lashes from side to side
- Hissing, spitting and growling

Submissive

- Dilated pupils
- Head held low
- Ears flattened
- Body crouched low to ground, almost shrinking in size
- Legs and feet tucked beneath body
- Tail pulled in tight or tucked between legs

The submissive cat

Scent Marking and Communication

Scent marking is the most important way in which cats communicate nonverbally, but it's the one we're probably least aware of because a cat's sense of smell is considerably different from ours, and cats can smell a host of things that we cannot detect. When we become aware of a cat's scent marks, it's usually because they involve litter box lapses.

Once a territory has been staked out, cats perform the ritual of scent marking to define it and reveal their presence to other cats. Cats scent mark their territories in several ways: by rubbing against things, scratching wood or objects, urine spraying and fecal marking. The primary way they mark their territory is by depositing pheromones secreted from glands on the chin and forehead, around the tail region and on the pads of the feet, which produce scents that are important in many aspects of feline social behavior. (Pheromones are chemical markers secreted by an individual that serve as an olfactory stimulus to other individuals of the same species for one or more behavioral responses.) Pheromones are also found in urine and feces. In fact, researchers in France have identified more than 17 substances that act as pheromones in cats. You can read more about the calming effect of pheromones to discourage urine marking and other stress-related behaviors in chapter 5.

Scent marking by rubbing

Scent marking by scratching

Scent Marking With Urine

Urine marking is an extremely important part of feline communication.
When sexually intact male cats (called tomcats) reach maturity at
around eight or nine months of age, they become possessive about
their outdoor and indoor territory and begin to spray urine to mark
it out. It is intact male cats who primarily do this, although intact
females and even neutered cats may also spray, especially when they
are threatened or stressed. The pheromones found in cat urine send
a strong message to other cats in the vicinity that a certain territory
is already occupied. Spraying is also a common behavior during mating season, when both males and females advertise their availability
via their pheromones.

Outdoor tomcats spend much of their time marking the boundaries of their home territory by spraying objects such as trees,
shrubs, fences, mailboxes, porches, the sides of houses, automobile
tires and other upright surfaces. The cat backs up to the object,
holds his tail erect with its tip quivering and then sprays a steam of
strong-smelling urine at a height suitable for sniffing (the penis
points backward except during sex). A tomcat's urine contains
pheromones and a fatty ingredient that leaves an unpleasant odor
that endures through even the most inclement weather. The urine

mark communicates the tom's presence to other males in the area. It is the male's calling card, announcing, "Keep out! This is my territory!" When an intruder wanders into another male's area, he can tell how recently the marking occurred—which is the main reason tomcats make daily rounds of their territory to deposit fresh markings.

While we're on the subject, research has found that cats who live as part of a community can determine the difference between the scent of urine sprayed by a cat who is a member of the group and one who is not. Spraying increases during the breeding season, possibly because the pheromones are helpful in attracting sexually receptive females to a male's territory.

Intact males who live indoors will more than likely spray household objects. The surface sprayed may be your furniture, table legs, curtains or walls. Some intact females spray around the house as well.

In addition to territorial scent marking, spraying may be triggered in any cat (intact or neutered) when a cat feels displaced or pushed aside, such as when a new baby or pet comes into the household, by changes in the environment, when the litter box is dirty or used by too many cats, when multiple cats are having social disputes, when a cat can smell or see another cat outside the house and even when a cat is not getting enough attention. These problems are discussed in greater length in chapter 5.

Scent marking by spraying urine

Scent Marking With Feces

Marking with feces is much less common than marking with urine. It most often occurs with cats who spend all or part of their time outdoors and with cats who once were strays. Feces often serve as an indication of rank. Dominant outdoor cats leave their feces in uncovered places as an impressive visual and scent signal. Most indoor cats will bury their feces in the litter box. However, a dominant male will sometimes leave his waste exposed if he tends to be aggressive or feels the need to define his territory. When cats use fecal marking to indicate their annoyance with something, they usually deposit the feces close to the problem area.

Keeping Your Cat Healthy

The aim of this book is to help you train your cats to use a litter box and, when necessary, to understand how to solve any inappropriate urination and defecation problems. Cats are instinctively fastidious in their bathroom habits; training the average cat to use a litter box is easy because felines have a natural inclination to bury their urine and feces. *Cats do not normally urinate or defecate outside their boxes unless something is wrong.* Refusal to use the litter box is usually the result of emotional or environmental stress, or medical conditions that prevent them from controlling their bladder or bowels.

But before litter box training begins—*indeed, before you even bring a cat into your home*—there are certain responsibilities you must accept. The most important is to keep your cat healthy. Cats provide all of us with companionship and unconditional love. In addition, they make jolly companions for children and adults and, according to scientific studies, have a therapeutic effect on the elderly, the lonely, the isolated, the troubled and the ill. We, in return, must be willing to provide love, attention, nutritious food, clean living quarters and bathroom facilities, immunity from certain infectious diseases and regular veterinary care.

When a cat suddenly starts to soil the house inappropriately or to spray urine, the first step is *always* to take the cat to the veterinarian for an examination to rule out medical causes. Feline lower urinary tract disease (FLUTD), bladder infections, bladder stones, kidney disease, cystitis, diabetes, thyroid problems, arthritis, internal parasites, senility and other conditions can all cause a cat to urinate or defecate inappropriately or to urine spray. When a cat has a medical problem, she may associate pain with the litter box and refuse to use it. Only after your veterinarian rules out any medical reason for house-soiling or urine spraying can you begin to investigate environmental or behavior problems.

From the moment a kitten or adult cat enters your home, his physical health depends on two people: you and your veterinarian. Have your veterinarian give your cat a thorough physical examination every year. Cats mature almost five times as fast as humans do, and more and more veterinarians are recommending even more frequent health checks (every 8 to 10 months) on the theory that early detection and treatment of problems can hasten recovery and prolong life. Your veterinarian will also advise which vaccinations are necessary for your cat's health and lifestyle.

It is your responsibility, though, to recognize warning signs of illness and report them quickly to your veterinarian. Your cat can't take himself to the veterinarian, so his wellness is in your hands. Early diagnosis and treatment of disease can prevent a great deal of pain and suffering. Help your veterinarian by using your senses of touch, sight and smell to know the signs of good feline health.

The Signs of a Healthy Cat

Every cat is a distinct being with unique characteristics that distinguish him from all others. Once you learn what is normal for your cat, any sudden changes in general appearance, behavior, play habits, appetite, temperature and other factors that accompany illness will be clearly apparent.

- Your cat should be in good general condition, with a sinewy and smoothly muscled body. He should be active, alert and lively, with a keen appetite and an energy level that's normal for him.
- Movement should be graceful, smooth and agile, with no limping, stumbling, stiffness or dragging of limbs.
- Although cats' coats vary in density and length, every coat should be sleek, glossy (dark colors usually are more glossy than light ones) and unbroken, with no bare patches. Longhaired cats should not be matted and tangled. The cat should lick and groom his fur regularly.

- The skin should be smooth and supple. Its color may range from pale pink or silver to brown or black. It may be non-pigmented or pigmented (normal in some cats with spotted, striped or blotched coat patterns). There should be no evidence of fleas, mites, dandruff, crusts, lesions, pustules or any other infestation or infection. There should be no masses, lumps or bumps on or under the skin, especially around the nipples.
- The eyes should be clear and shining, with no excessive tearing, mucus discharge or sensitivity to light. The third eyelid (also called the haw or nictitating membrane), at the corner of each eye near the bridge of the nose, should be almost invisible. Unlike dogs, cats usually do not have eyelashes. The skin folds of the eyelids should be smooth. Roll down the bottom lid with your thumb to examine the lining of the eye. It should be pink, not ashy or bright red.
- The skin on the external earflaps and inside the ears should be pale pink. Brilliant pink, red or brownish skin indicates trouble. Smell the ears: healthy ones smell clean, while unhealthy ears smell foul. A little wax is normal; excessive amounts of dark wax are not. There should be no pawing of the ears or frequent shaking of the head.
- The nose should be cool and clean with no sticky or yellowish secretions.
- Examine your cat's teeth regularly. The insides of the upper incisors should touch the outsides of the lower incisors when the mouth is closed. An overshot bite (upper front teeth extending beyond lower front teeth) or an undershot bite (lower front teeth extending beyond upper front teeth) is abnormal. Teeth should be firmly implanted and not loose. They should be white in a kitten but may yellow slightly with age. There should be no tartar buildup on the teeth or around the gum line. The gums and tissue inside the mouth should be pink—never red, inflamed, bluish or ashy. The breath should smell pleasant.
- Body openings should function properly with no abnormal discharges. Membranes should be clean, smooth and pink.

- Become familiar with your cat's urinary patterns because changes may indicate changes in body chemistry. Urine should be clear yellow with a characteristic aroma, never orangey. Your cat should go into the box, do his business, scratch around and get out.
- Diet influences the volume, color and odor of a cat's feces, but the stool should be well-formed, typically brown, and it should be eliminated regularly. Stools should not be loose, strangely colored, blood-streaked or putrid smelling.
- Breathing should be clear and regular, with no wheezing.
- Normal rectal temperature for a cat is between 100.5°F to 102.5°F (38°C to 39°C). Excitement, exercise or heat can temporarily cause the temperature to rise.

The Signs of a Sick Cat

It's not difficult to know when your cat is sick. Any deviation from normal good health will usually be accompanied by one or more changes in body functions and behavior. As an observant owner, you will be able to immediately recognize these changes and report them to the veterinarian—an invaluable help in the discovery and diagnosis of early illness. These are the most common signs of illness, but remember that any deviation from your cat's normal behavior is a cause for concern.

- Changes in appetite or fluid intake
- Excessive or rapid weight gain or loss
- Fever
- Behavioral changes: apathy, depression, listlessness or viciousness
- Limping or lameness
- Bloody, frequent or uncontrollable bowel movements
- Constipation or any other problem in passing feces normally
- Orangey, cloudy or blood-tinged urine
- Constant straining to urinate or inability to urinate
- Frequent trips to the litter box, often accompanied by crying, and frequent licking of the genitals

- Abdominal swelling or tenderness, lumps beneath the skin or on the breasts
- Abnormal discharge from any body opening
- Prolonged vomiting or attempts to vomit
- Prolonged coughing or sneezing
- Strained breathing or wheezing
- Sticky, yellowish or greenish nasal discharge or crusting on the nose
- Partial covering of each eye by the third eyelid or discharge from the eyes
- Ear problems: foul odor, excessive wax, excessive shaking of the head
- Foul breath
- Excessive salivation
- Intense biting and scratching of the skin and coat
- Unkempt or matted coat or any other sign that your cat has stopped grooming himself
- Hair loss, baldness, sores, pustules, lesions, excessive external parasite infestation or any other skin irregularity

Feline Lower Urinary Tract Disease

One of the more common reasons for litter box lapses is feline lower urinary tract disease (FLUTD—formerly known as feline urological syndrome, or FUS). FLUTD actually refers to a complex series of diseases with multiple causes that affect the lower urinary tract of cats. It can range from a painful irritation of the bladder to a life-threatening condition, and signs of the disease should never be ignored. Because cats with FLUTD often associate their litter box with the pain they experience upon urination, they frequently bypass the box and look for other places in which to urinate, hoping things won't hurt so much in a different location.

A cat's urinary system works very much like our own. The kidneys filter waste products from the blood and funnel the liquid waste through narrow tubes called ureters that empty into the bladder. Each kidney has its own ureter. As the urine drains into the bladder, the bladder expands like a water balloon. When the bladder becomes full and distended, nerves send impulses to the brain and the brain sends back impulses causing the bladder to contract. The cat then urinates through a third tube called the urethra, a passageway that goes from the bladder out through the vagina or penis.

FLUTD is most common in cats from one to six years old, but it can affect both male and female cats of all ages. The condition is more serious in males because of anatomical differences: a female cat has a short and wide urethra, while a male has a long and narrow urethra. Although FLUTD can be caused by a bladder irritation, bladder stones, bacterial or viral infections or urethral blockage, the last—blockage—is the most serious. Small stones or sand form in the urine and plug the urethra, making it impossible for the cat to empty his bladder. Because males have a more narrow urethra, they are more prone to this type of blockage.

As the urine backs up, the bladder expands and becomes painfully full, and the cat will exhibit the symptoms described below. If the cat cannot urinate at all, pressure pushes backward from the bladder through the ureters and into the kidneys. In this case, the urine backs up to the kidneys, the kidneys become damaged and may stop working altogether and toxins build up in the blood. At this point, without immediate veterinary treatment a cat can very quickly die.

Most cats with FLUTD exhibit at least some of the following signs, to varying degrees:

- Frequent urination
- Frequent trips in and out of the litter box
- Straining to urinate, although the amount of urine passed during each attempt is small
- Crying from painful urination
- Excessive licking of genital areas
- Bypassing the litter box to urinate on cool, smooth surfaces such as bathtubs, sinks, tile floors and similar locations
- Blood in the urine, often accompanied by a strong ammonia odor
- Listlessness, poor appetite, excessive thirst

Because FLUTD can be life-threatening, even for female cats, you must be able to recognize these signs when they occur and seek immediate veterinary treatment. When diagnosed early, some forms of FLUTD are curable, while others can be controlled with long-term medical and dietary management.

Intestinal Parasites

Diarrhea and defecating outside the litter box can be signs of internal parasites and other gastrointestinal upsets. Intestinal parasites are common ailments of cats—even indoor cats—and the most responsible and meticulous owners cannot always shield their cats from infestation. Parasites seem to be most common in kittens and young cats, although cats of any age can be infested. Some of the most common kinds of internal parasites are roundworms, hookworms, tapeworms and whipworms. Other internal parasites that can plague cats include the protozoans *Giardia*, *Coccidia* and *Toxoplasma gondii*.

The signs of a parasite infestation include intermittent, foul-smelling (often bloody) diarrhea, mucus in the stool, a potbellied appearance, weight loss and, with tapeworm, ricelike debris or moving segments sticking to the cat's anal area or in the litter box.

All internal parasites should be specifically identified by laboratory analysis (in most cases by microscopic examination of a fresh stool sample) and treated by a veterinarian. Once the type of internal parasite is identified, your vet will dispense the correct medication for worming, calculating the dosage based partly on your cat's age and physical condition.

Worming medicines are formulated to kill internal parasites and can be dangerous when administered in excessive amounts, too frequently or for the wrong species of parasite. That's why it's important to take your cat to the veterinarian, rather than relying on an over-the-counter dewormer that may or may not be for the type of parasite your cat has—and that offers only general dosing directions that do not take your cat's unique health circumstances into account. See your veterinarian, and follow his or her advice to the letter.

Rapid or Frequent Dietary Change

Cats are content to eat the same diet day after day. They like to eat in the same location, at the same time and from the same clean dish every day. It may sound boring to you, but it's reassuring to them. Variety can be a good thing for cats, but not all cats can eat everything. Many cats are lactose-intolerant, for example, and dairy foods can cause diarrhea.

Rapid or frequent dietary changes can cause gastrointestinal problems in some cats. If it becomes necessary to change your cat's food, do it gradually, over a period of a week to 10 days. Dietary changes due to medical reasons should always be supervised by a veterinarian.

A Word About Toxoplasmosis

Toxoplasmosis—infection of the parasite *toxoplasma gondii*—is of special concern to pregnant women and to people with compromised immune systems, such as those with AIDS, organ transplants or cancer. It can be spread by handling or eating raw meat, by drinking unpasteurized milk or eating unwashed fruits or vegetables, by gardening in infected soil and through contact with infected fecal material. Cats are one of the hosts that can carry and pass this parasite to humans through infectious cysts shed in their stool.

Toxoplasmosis is one of the most common human infections throughout the world. The Centers for Disease Control and Prevention (CDC) estimates that more than 60 million people in the United States probably harbor the parasite, "but very few have symptoms because the immune system usually keeps the parasite from causing illness." While an active toxoplasmosis infection generally causes mild flulike symptoms in humans that go away on their own in a few days, it is very dangerous to fetuses in the first trimester of pregnancy.

In the past few years, an alarming number of comments and articles about toxoplasmosis have appeared in women's magazines that have been especially disparaging to cats. The most common, unquestionably, is the old wives' tale that it is unsafe for pregnant women to keep cats because touching them or being around their litter boxes will cause their babies to be physically or mentally deformed. But although cats have been blamed for infecting their owners with this parasite, the reality is that *you are far more likely to contract toxoplasmosis from infected meat or soil than from your cat.* Cats and pregnant women can safely coexist—and have been doing so for thousands of years.

Studies show that the major source of toxoplasma infection in humans is eating raw or undercooked meats and *not* contact with cats. Cats who live indoors, who have never caught and eaten birds, rodents or other small animals and have never been fed raw meat are extremely unlikely to be infected. On the other hand, most humans handle raw meat frequently and therefore have had some exposure. And once you have been infected and your immune system has fought off the infection, you are immune for life—even if you become pregnant. Your doctor can do a blood test to check for prior toxoplasma exposure.

If you have never been exposed, a very unlikely chain of events would have to occur for you to get toxoplasmosis from your cat. Cats shed toxoplasma cysts (the egg stage) in their feces 3 to 10 days after eating infected tissues. They will shed the cysts for up to 14 days, and afterward it is unlikely that they will ever shed them again—even after repeated exposure. So only a recent infection is contagious. Even then, toxoplasmosis is only infectious under a particular set of extremely unlikely circumstances. The cysts do not become infectious to other animals and humans until one to four days have passed, so the infected stool has to have been in the litter box for more than 24 hours. Then, you'd have to touch the stool with your bare hands and then get some of the stool in your mouth or eye—which means touching them after you have scooped the litter box but before you have washed your hands. This is not something *anyone* should be doing—much less a pregnant woman!

A few simple precautions will keep you safe from any likelihood, no matter how remote, of getting toxoplasmosis from your cat.

- Have someone else in your family scoop and clean the litter box during your first trimester of pregnancy.
- If you must scoop and clean the box, wear rubber or plastic gloves when you do and wash your hands with soap and water afterward.
- Scoop the litter box daily to prevent any parasites from becoming infective.
- Dispose of any cat feces in a sealed plastic bag or flush them down the toilet.
- Wash the litter box with scalding water. It's not necessary to use a disinfectant, because they have no effect on toxoplasma cysts.
- Feed your cat commercial cat food or cooked homemade food, never anything raw or undercooked.
- Keep your cat indoors to prevent any accidental encounter with birds or small animals. Don't let your cat kill and eat birds or rodents.

For the far more likely scenario of getting toxoplasmosis from raw meat or garden soil, take these precautions.

- Do *not* eat raw or undercooked meat, drink unpasteurized dairy products from any animal or eat unwashed vegetables or fruits that could be contaminated with soil containing toxo-plasma cysts.
- Wash your hands, as well as utensils and cutting boards, with soapy water after handling raw meat.
- Cook meat thoroughly before it is eaten. The CDC advises that the internal temperature of meat should reach 160°F.
- Wear gloves while gardening or cleaning sandboxes to avoid coming into contact with contaminated soil. Wash your hands thoroughly after working in the garden.
- Place a covering on your sandbox to prevent neighborhood cats from using it as a litter box.
- Avoid stray cats, especially kittens.

3

The Scoop on Litter, Litter Boxes and Accessories

A recent market study, published by the industry trade magazine *Pets International*, reports that cat litter sales in the United States have reached $1.3 billion a year! That's nearly 500 billion pounds of litter. The popularity of cats as pets has skyrocketed not only in the United States but also around the world. Euro-Monitor International, publishers of market research reports, estimates that the global cat population exceeds 200 million and global cat litter sales have topped $2.71 billion.

How things have changed! Years ago, cats lived most of their lives outdoors. Litter and litter boxes were unimportant. When nature called, cats urinated and defecated whenever and wherever they pleased, usually in garden soil or sand. Even when they were allowed to come inside, cats had to get by with rudimentary boxes filled with soil, sand, sawdust or wood ashes. None of these materials was very functional—they were messy, they were unbearably smelly and cats tracked them through the house.

In the late 1940s, Edward Lowe, a Michigan building supply company salesman, conceived the idea of using granulated clay as filler for litter boxes. The substance worked so well that Lowe started packaging it, and Kitty Litter became the first consumer product of its kind in the United States. Clay was definitely more absorbent than soil, sand, sawdust or ashes. But although solid wastes could be easily scooped, the litter did become saturated with liquid waste within several days and had to be emptied out and replaced completely.

Over the next four decades, the litter industry changed very little until another milestone in 1984 when Thomas Nelson, Ph.D., a biochemist, developed the first clumping litter—superfine, sandy grains that absorb 12 to 15 times their weight of any liquid. When a cat urinates on them, the grains expand and stick to form clumps that can be scooped up and discarded, leaving clean, odor-free litter that can be changed less often than traditional clay litter. When clumping litters were first introduced, their sales amounted to less than 1 percent of total litter sales. Today, clumping litters have become so popular that they represented an amazing 55 percent of sales in the entire cat litter industry in 2003.

Today, several manufacturers have also marketed natural or alternative litters for people looking for healthier or more ecological alternatives for themselves or their cats. These are made of eco-friendly materials such as cedar shavings, pelleted wheat and wheat grass, corn cobs, processed orange peels, peanut shell meal treated with herbs, cedar chips, silica gel beads and crystals and even recycled newspapers. Alternative litters are said to be more absorbent than clay and control odors more efficiently. Some cats, however, don't like the feel of the pellets under their paws. And most of these litters (although not all) do not clump.

Litter box design has also progressed. There are a host of new and appealing items on the market that include standard rectangular plastic boxes (with and without rims), hooded boxes, self-sifting boxes and automated systems that scoop up after your cat urinates or defecates so that you don't have to.

There is such a vast array of items to choose from that it can be confusing to figure out what will best suit your cat's age, size and general health, and your home and your lifestyle. What type and size of litter box best suits your cat? What type of litter is best? How much litter should the box contain? How many boxes do you need? You also need to think about what make sense for you. How often are you willing to empty and scrub the box? How often can you scoop?

However, to make sure your cat uses his litter box regularly, bear in mind that the primary consideration is what best meets the needs of your cat—the litter box and litter that he prefers, *not you*. *You can*

How Cats Eliminate

Let's learn a little more about the elimination behavior of cats. According to behaviorists Daniel Q. Estep, Ph.D., and Suzanne Hetts, Ph.D., the elimination process normally starts with the cat sniffing an area and then scratching at the surface with a front paw, as if she were digging a hole. The cat moves around in a circle, squats and urinates or defecates on the scratched surface.

When the cat is finished, she will stand, turn, smell the soiled area and scratch the litter again in an attempt to bury the waste. "Some cats will repeat the sniff and scratch sequence more than once before walking away," say Estep and Hetts. "There are great differences in how much different cats can scratch. Some will take only a swipe or two and never really bury their waste, others may dig as if they are building sand castles. Despite this normal variation, most cats will show some pawing at the surface unless something is wrong, like an aversion to the surface or area."

never train a cat to use a box or litter that he finds unpleasant or uncomfortable. It's just not possible. If you try, the inevitable result will be that your cat refuses to use the box.

You must also remember that no matter how "self-cleaning" a box or a litter claims to be, there is no getting around the fact that sometimes you will have to completely empty and scrub out the litter box. Cats will not use a dirty box—and really, can you blame them?

Many product brand names are mentioned in this chapter—popular items made by reliable manufacturers that are widely available at pet supply stores throughout the country. I've done this to give you an idea of the multitude of items that are available. I am not employed by and do not act as a consultant to any of these manufacturers.

Litter Box Choices

A litter box is simply a container that holds your cat's absorbent litter or substrate material. All litter boxes used to look alike: plain, shallow, rectangular pans made of plastic. Today's manufacturers not only make litter boxes in different sizes, shapes and colors, they have also marketed complete systems designed to hygienically do away with the dirty work of litter box care. These new systems fall into two main types: sifting boxes with slotted or screened sections that separate waste, and automated or self-cleaning boxes that automatically rake through soiled litter and deposit it into a disposable unit or bag. Technology has advanced so that, for those who can afford it, the job of scooping litter to remove cat waste is fast going out of style. Here's a brief review of the different types of litter boxes you can choose from.

Basic Litter Box

The basic litter box is a rather shallow, rectangular-shaped pan made of heavy-duty, stain- and odor-resistant plastic that is easy to wash and disinfect. The average box is about 12 inches wide by 18 inches long, and about 5 to 6 inches high. Smaller sizes are available for kittens, and you can get extra-large ones for big cats or multiple-cat households.

Size is important because you need a litter box that's big enough for your cat to turn around in and deep enough to prevent spillage. Although this most basic design is inexpensive, it's very practical if you don't want to spend a lot of money. If you prefer cylindrical shapes, Booda makes a 17-inch round, pearlescent and metallic style, called the Litter Bowl, with matching litter mat.

Basic litter box

Basic Rimmed Litter Box

This is another shallow, rectangular box that is made of heavy-duty plastic. It's also known as a framed litter box. The only difference between this box and the basic litter box I just described is that this box comes with a three- to four-inch plastic rim that curves inward and fits around the edges. The rim is designed to prevent litter from being scattered outside the box. It, too, is inexpensive, and is especially practical if you have a cat who is determined to scratch and dig.

Rimmed litter box with a plastic liner

Hooded Litter Box

This litter box has a hood or cover that snaps or locks onto the basic rectangular bottom. It looks so much like a tiny house that kittens have been known to crawl inside and go to sleep. Hooded boxes come in assorted colors, in large and jumbo sizes and, for multiple-cat households, in extra-giant sizes. Some of the more advanced models have odor-absorbing filters in the hood vent to reduce odor and keep air fresh. However, you must remember that while the filter will help keep the air in your room fresh, it will not reduce odor *inside the box*. A dirty box will still smell awful to your cat.

The Boodabox Swingtop has a hinged lid that enables you to scoop litter without completely removing the top. You just tilt back the lid and then snap it back into place when you're done. Another hooded

Hooded litter box

model is the Catty Corner Litter Box, with a swinging hinge door, which is triangular and can be tucked into any out-of-the-way corner in a home. Another innovative box with a privacy cover comes from Aspen Pet Products. It's a large-capacity pan for multiple-cat households and has a push-button retractable hood that allows access to cats but not dogs or children.

Hooded litter boxes provide privacy, help contain odors and urine spray, isolate solid waste from children and other pets and prevent a cat from scattering litter outside the box. Should you consider buying one, be sure there is enough headroom for your cat, that the hood is not so flimsy that it can be pushed off and that it snaps securely to the bottom.

Sifting Litter Box

A sifting litter box is similar to the basic rectangular box, but it comes in three parts: two bottom trays and a fine mesh sifter that sits on top. As you lift the sifter, any solid waste and urine clumps stay inside for disposal, while the unsoiled litter falls back into the bottom. There's no need to use a scooper.

The most popular styles are the Lift 'N Sift and Quicksand Cat Litter Sifting System. Both work well with traditional clay, clumping and silica gel litters. With both, when the bottom two trays are nestled on top of each other, the slots extend over each other to form a solid bottom. The third (top) tray becomes the sifter. To clean the box, you simply lift the top tray containing the solid waste straight up, and the clean litter sifts through the slots to the sections below. Once the tray is emptied and rinsed off, you place it at the bottom of the stack, beneath the other two.

Sifting litter box

Another type of sifting box is the Litter Glitter. It has a rectangular tray with a large diamond-shaped screened scoop that sits inside the litter box or beneath it. When you lift the screen, the waste separates from the clean litter, and you dispose of it and slide the screen back into place.

Automated or Self-Cleaning Box

Automated or self-cleaning litter boxes are the most high-tech systems you can buy. They work best with clumping litter. These range from devices that roll over and capture the waste clumps to fully automated appliances with electric-eye sensors. A popular roll-over box is the EverClean Self-Scooping Litter Box, which has a snap-on lid that contains a scooping screen. You roll the box over once, the screen traps the waste clumps, then you open the lid and empty the waste. Another popular roll-over model is the Omega Paw Self-Cleaning Litter Box with hood. You remove waste from the litter by rolling the top of the box 180 degrees to the floor and then back. The waste goes into a removable tray for disposal, and the clean litter falls back into the box.

There are also deluxe systems that barely require you to do anything—they usually work with household current or eight D batteries. Two such products are LitterMaid and Petmate Purrforma Plus; they are expensive but fully automatic and do almost all the cleaning for you. You fill the LitterMaid box with clumping litter and your cat uses it just like a normal litter box. Approximately 10 minutes after the cat uses the box, an infrared sensor senses that the animal has exited the box and signals an automatic sifting comb to rake through the litter and scoop up any waste. The waste is deposited into a sealed, airtight container. Meanwhile, the comb glides smoothly over the litter, returns to its original position and is ready to repeat the cycle.

The Petmate Purrforma Plus is a jumbo hooded automated system that also uses state-of-the-art infrared sensors to detect when a cat enters and leaves the unit. You fill the box with clumping litter

Self-cleaning litter box

and, after an adjustable 10-, 15- or 20-minute delay, Purrforma quietly rakes waste into a large-capacity box, sealing away clumps in a disposable bag. You simply remove the bag and drop it in the trash. The box comes with a paw-cleaning ramp that helps minimize tracking litter onto the floor.

Disposable Litter Box

This is a folding box, often made of waxed or crimped cardboard, which is disposable and recyclable. Disposable boxes are too unstable to use every day, but they are included here because they are especially useful at cat shows, for traveling, for vacations and for cats who are ill for a short time (particularly in a multiple-cat household, where you don't want to risk spreading any illness). After the show or the vacation, or when the cat's health has improved, you just throw away the box.

What Type of Litter Box Is Best?

Whether your puss is strictly an indoor cat or goes outdoors occasionally, a litter box is the most important item you will provide for him. The box you choose should be made of non-absorbent, stain- and odor-resistant material—plastic, more than likely—that is easy to clean with mild detergent, such as a dishwashing liquid, and warm water.

Consider the size of your cat when choosing the litter box. Kittens need a small, low-sided box because they are too tiny to climb into a larger one. Veterinarians often recommend using shallow aluminum baking pans for kittens under 12 weeks of age until they are able to get in and out of conventional litter boxes. As the kitten matures, you can get a larger one.

For an adult cat, pick a litter box that will be roomy enough for him to urinate and defecate in several areas and still have some clean spots to stand on. You don't want the box to be so small that the cat's rear end hangs over the side and misses the litter when he eliminates. A large cat with a big body (such as a Maine Coon) who likes to turn around and around to find the "right spot" will, of course, need a bigger box than a small cat who hops into the box, squats, elimi-nates and hops out without making a big production of it.

Whatever size you choose for your adult cat, the bottom of the litter box should be deep enough (four to five inches) to give him enough burying room without spilling the litter outside.

Like kittens, older cats may need a box with lower sides. Any loss of mobility, such as arthritis or injury, can make it difficult for your old friend to get into and out of the box. If your older cat is sud-denly "missing" the box, try providing one with lower sides.

If you're considering an unusually shaped or high-tech box, bear in mind that while some cats willingly accept them, other cats find them unacceptable. The sound and movement within the box, even if it starts up long after the cat has exited, can make some cats ner-vous about getting into the box. If you want to introduce a high-tech box, *always* give your cat a low-tech alternative, just in case.

Open or Covered?

Some cats like the privacy and security of a covered litter box, while others prefer an open box where they can observe everything that's going on in the area. You may decide to experiment at first and pro-vide both types to learn which your cat prefers. Here are some points to consider:

- A cat who is especially shy or reserved may prefer a covered box.
- Hooded boxes are good choices for cats who like to dig and kick a lot of litter outside the box.
- A covered box can be helpful in isolating your cat's feces from any resident dogs. (Some dogs enjoy eating cat feces because they are attracted to its high protein content. Repulsive as it may sound, doing so poses no health hazard to your dog.)
- The hoods on the box tend to reduce air circulation. Soiled litter takes longer to dry, creating a perfect environment for odor-causing bacteria to grow.
- The reduced air circulation can also be aggravating for cats with asthma and other chronic respiratory ailments.

- Cats who object to covered boxes can *sometimes* be retrained. Simply take off the lid and let the cat use the bottom tray until he gets used to it and then replace the lid. But remember, you cannot force your cat to use a box he dislikes.

If you choose a covered box, make sure the opening is large enough for your cat to enter and exit comfortably. The box also should have a deep bottom and be roomy enough for your cat to turn around, scratch, dig and position himself comfortably before he eliminates.

One more thing to consider: in a multiple-cat household, a covered box may make it easy for an aggressive cat to lie in wait and ambush a timid one as he leaves the box. "Multiple-cat households often have a bully cat or cats," writes Marty Becker, D.V.M., and Janice Willard, D.V.M., in *Cat Fancy* magazine. "Humans sometimes have a hard time seeing fault in a beloved pet, but in all honesty, some cats can be dreadful bullies. They lie in wait for their feline victim in places they know the cat will need . . . like the litter box."

Litter Choices

Litters these days can be non-clumping, clumping, pelleted, flaked, granulated, dust-free, dust-reduced, chemical-free, deodorizing, scented, unscented or low-tracking, and some are even treated with catnip! Just when you think there's no way to create more kinds of cat litter, someone comes along to do it. With so many choices facing today's cat owners, it is no surprise that we are often confused about what to buy.

Some advice before we begin: Check with the breeder, owner or shelter from which you acquired your kitten or adult cat and ask what brand and type of litter the animal is used to. Then, *don't change it*. Continue to use that same litter at your home. Don't buy something else because it's on sale. Switching litters suddenly, especially with a cat who must adjust to a new home, can cause confusion and elimination problems.

Also, tempting as it may be, *do not* use fragrantly scented additives to camouflage litter box odors. What smells pleasant to you may be a complete turn-off to your cat. This is especially important when you have a new cat in the house. You want your new kitten or adult cat to find his litter box easily, and a tiny whiff of his own urine or feces will remind him that the litter box is the right place to eliminate.

Because cats are sensitive animals with different temperaments and preferences, there is no single litter that is right for every cat. What pleases one cat may be objectionable to another. You must choose the right type *for your cat*—not for you—because cats who aren't satisfied with their litter may stop using their litter box. No matter how much you love that new recyclable, flushable litter, if your cat doesn't like it, he will stop using the litter facilities you provide and start urinating or defecating on some other surface.

Here's a brief review of the different types of litter you can choose from.

Traditional Clay Litter

White or natural clay is the most basic type of litter and, since being introduced in the late 1940s, has been the foundation of the cat litter industry. Clay litter is inexpensive and economical because it is widely available as a natural resource throughout the United States. Basically, clay is strip- or open-pit-mined from the ground, fragmented into small particles that are appealing to cats' paws and then dried. Traditional clay litters come in a variety of formulas, including unscented, scented, antibacterial, multi-cat and low-tracking.

Traditional clay litters absorb fluids efficiently, but because they do not form clumps, urine tends to pool in the bottom of the litter box. That means the entire contents of the box *must* be dumped out frequently and replaced with fresh litter to keep the box from becoming soggy and smelly.

Clay litters are quite dusty; the finer the particles, the easier they are to track outside the box. They also tend to cloud upward while being poured into the litter box, which can be irritating to allergic or asthmatic humans and cats.

Some popular brands include KleenKitty, Fresh Step, Cat's Pride, C-9 Premium, JonnyCat, Tidy Cat, Kitty Litter, Kitty Litter MAXX, Sani-Cat and Hartz Mountain.

Clumping Clay Litter

Most clumping litters are made from either sodium bentonite or atta-pulgite—two highly absorbent types of clay. Sodium bentonite is more commonly used because it's relatively inexpensive. When flushed, however, sodium bentonite clumps tend to swell and clog pipes, so you don't want to dispose of them in the toilet—even though some manufacturers claim their products are flushable. Attapulgite costs more, but it breaks apart in water and is more toilet-friendly. *Do not flush any kind of clay litter if you have a septic system.*

The clay is ground into very small particles that look like fine-ly textured sand. When cats urinate on clumping litter, the sandy granules swell, absorbing more than a dozen times their weight and moisture, and form hard clumps. Both liquid and solid waste can be easily scooped from the litter box, while the rest of the litter stays relatively dry. Clumping litters are very economical and efficient in multiple-cat households; it's easier to keep the box clean day to day, and you don't need to change the entire contents of the litter box as often as with non-clumping clay.

Cats love clumping litters, probably because they are the closest in texture to sand or soft dirt. There are several kinds of clumping litters: unscented, scented, no- and low-tracking, extra-strength (for multiple cats) and heavy-duty formulas (for multiple cats and cats who like to dig). The extra-strength and heavy-duty types form clumps so solid that they don't even break when dropped.

Dust can be a major problem with clumping litters because the granules are so fine. Tracking is also a problem; the sandy granules tend to stick to the paws when cats exit their boxes, and conse-quently, they can leave trails of litter around the house. Since their introduction, clumping litters have been refined and modified many times.

Popular brands include: Cat Attract, Better Way, Ultra Clump; Ever Clean (there's a formula for every situation: ES/Extra Strength, HD/Heavy Duty, LT/Less Tracking, and FG, scented with Field Guard, a germ fighter that emits a mild fragrance); Scoop-Lite, Jonny Cat Scoop, Tidy Scoop and Arm and Hammer.

Natural and Alternative Litters

These litters are generally more expensive than clumping and non-clumping clay. They come in pellet, flake and granule form and are made of a variety of substances. They tend to absorb moisture and control odors better than clay. Most are biodegradable, contain no chemical additives, generate less dust and cause little or no tracking. Most types, according to manufacturers, can be flushed or composted. However, composting cat litter is *not* a good idea because of the risk of spreading parasites and diseases.

There can be some problems, though, with alternative litters. Those that are ground into flakes and granules are soft and light in texture—actually lighter than traditional clay litter. This appeals to cats but tends to cause tracking. The other problem concerns the size of some litters, which are made from substances that are pressed into pellets. Research has shown that cats prefer fine litter that feels like earth or sand under their paws, and although larger pellets tend to eliminate tracking, some cats don't want to use them.

Let's take a look at the different types of alternative litters.

Paper

Made from recycled paper or newspaper, paper litter comes in either pellets or small pieces—some so small that they look and feel like traditional clay. Both forms are lightweight and can be flushed down the toilet in small quantities. Paper litters are said to be up to 300 percent more absorbent than some clay litters. As the pellets absorb urine, they swell up to contain moisture and odor and form clumps that are easy to remove. They are non-toxic and have been treated so they will not smudge the cat's feet with ink.

Because their soft texture is kind to tender cat paws, veterinarians often recommend using paper litters for recently declawed or post-surgical cats. Paper litters are also helpful for geriatric cats or cats who suffer from allergies. Popular brands include Yesterday's News, Good Mews, ECOfresh, Bio-Flush and Kitty-Soft.

Plant Products
These litters come in both clumping and non-clumping forms. They are made of a wide variety of materials, such as wheat, grass (usually western or northern winter wheat grass), alfalfa, grain, corncobs, peanut hulls and citrus peels, which are formed into pellets or cereal-like granules. They have a natural, fresh smell that controls odors without chemical additives, and they are not harmful if ingested. Most are air-cleaned before being packaged, to eliminate dust.

Plant-based litters are highly absorbent; they soak up liquid like a sponge. Most plant and plant by-product litters can be flushed down the toilet in small quantities. Popular brands include SweatScoop (wheat), Heartland Wheat Litter, Wheat 'N Easy, Clump 'N Flush (wheat), Cat Country (grass), Double Fresh Green (grass), Cat Works (grain), FIELDfresh (corncob), Cobby Cat (corncob), World's Best Cat Litter (corn), Premium Litter Plus (peanut shells) and Citra Fresh (citrus peel).

Wood
These litters are formulated from various woods (including cedar, pine and aspen) and wood by-products, including sawdust and bark. They are lighter in weight than clay and come in shavings, chips and pellets of various sizes. Their appealing, woodsy scent and superior ammonia-absorbing properties help control litter box odors. *Some* wood litters can be flushed down the toilet in small quantities. They do tend to track easily. Popular brands include CareFRESH, Cedar Lite, Fresh Step Cedar Scooping, Gentle Touch, Pine Fresh, Mountain Cat, LitterLove Lite & Natural and C-9 Cedar.

Silica or Crystal Pellets
These are the hot new products in the cat litter industry. Made of biodegradable silica gel, they are small, lightweight, round beads or "pearls." Each bead is covered with micropores that capture and

absorb moisture and trap odors. Odor is immediately locked inside the beads and all the liquid is absorbed, leaving the outer layer of the pearls dry to the touch. When a cat urinates, you can literally hear crackling as the pearls absorb the liquid.

Silica pearls work so well that you won't find urine pooling in the bottom of the litter box. However, you still must scoop out feces daily. After about 30 days, or when the pearls turn yellow, they must be thrown away and the box scrubbed and refilled. Although these litters track less than other types, they do roll around the room like balls when they are kicked out of the box. A recently marketed new shape is said to reduce this tendency. Popular brands include Ultra Pearls, Tidy Cats Crystals, Litter Pearls, Litter Pearls Track-Less, Crystal Clear Litter Pearls, Exquisi-Cat Crystal and Pearl Fresh Cat Litter.

Medicinal Litters

There's also another type of litter—medicinal—for cats with special needs. These litters are available mostly from veterinarians, and they alert cat owners early on to symptoms of medical problems such as Feline lower urinary tract disease (FLUTD) and diabetes. Medicinal litters are coated with special chemical indicators that turn different colors in the presence of some substance in the urine.

For FLUTD, for example, when the urine has a high alkaline pH level, the litter changes color, usually red, on contact. There are also litters that change color to indicate the presence of blood in the urine—a sign of a range of problems, including FLUTD.

There are several different products for cats with diabetes. One is a special non-absorbing litter that lets you dip a test strip into the urine in the box to determine the presence of glucose. You can also get packets of powder that is sprinkled onto litter; the powder changes color to indicate the presence of blood or excessive sugar. There's even a litter system for diabetic cats in which the urine drops through one tray into another for more accurate glucose testing. Your veterinarian can provide more information about these and other medicinal litters.

Catpaper

While not a litter, Catpaper deserves a mention here. Odor-free and non-toxic, Catpaper is made from recycled absorbent paper bonded to a clear plastic backing. It's available in sheets for small areas and/or rolls you can cut to fit larger areas. Catpaper protects against all sorts of cat accidents by absorbing urine, while the plastic backing stops the liquid from soaking through. Use it in the litter box after de-clawing or other surgery, for diabetic cats, under bedding areas for cats who leak urine and as a disposable mat under the litter box for those near misses. For cats who spray, taping Catpaper to the wall reduces cleaning time. It also protects furniture from urine, hair and dander.

What Type of Litter Is Best?

Your choice of litter is as important as the selection of the litter box. New cat owners, however, can be frustrated when it comes to making a decision about which litter to use because the choices seem endless. The important thing to remember is that you must pick a litter *your cat likes*. And then you must stick with it because frequently changing litters is a sure way to get your cat to stop using the box.

For cats, the most important feature of litter is how it feels under their feet. Their natural choice is sand or soil. In a study conducted several years ago, animal behaviorist Dr. Peter Borchelt found that most cats prefer a fine-grained, sandy surface, similar to the type used in children's sandboxes. The more coarsely grained the litter, the less the cats liked it. Borchelt also found that cats disliked litters with a strong scent and litters that are very dusty.

Basically, the requirements for any litter you choose are the same:

- It should remind the cat of his natural environment.
- It should be absorbent.
- It should be fragrance-free.
- It should not be too dusty.

The kind of litter you choose is up to you. However, to make a kitten or cat more secure in new surroundings, *always* start with the same litter that the breeder, former owner or shelter used. If, for any reason, it becomes necessary to change, you should do it gradually, mixing the new litter in with the old a little at a time over several weeks, so the cat has time to get used to the new litter. Nevertheless, it is very important to keep in mind that cats are creatures of habit and *they don't like changes!*

Cats who have been surgically declawed can have very sensitive paws. They often prefer the soft, fine granules of clumping litters or pelletized newspaper.

Persian and other longhaired cats will sometimes refuse to use clumping litter because it can cling to their hair and paws. Be extremely vigilant with a longhaired cat: the litter could stick to the hair under the cat's tail or around the anus, and moisture could cause it to adhere to the coat and clump into a mass. The mass can harden and seal the anal opening, making it difficult for the cat to have bowel movements. Cats often try to remove the clumps by pulling out the soiled hair with their teeth. Check your longhaired cat often and keep the hair under the tail trimmed short.

Cats of all ages occasionally eat their litter. Unfortunately, cats have also been known to accidentally ingest litter during the process of licking their fur and paws when they are grooming. *Although there is no scientific literature on any relationship between eating of clumping litter and intestinal blockage or gastrointestinal upsets in kittens, most litter manufacturers suggest using clumping litters for kittens after they are eight weeks of age or older.* Consult your veterinarian if this is a concern.

How much litter do you need? Start by filling the litter box with approximately two inches of litter. Kittens need a little less. Tenacious diggers need a little more.

Litter Box Accessories

The only litter box accessory you absolutely need is a scooper. Everything else is optional. As always, your cat's preferences are top priority. Some of these products, such as litter box liners, are designed to help you with clean-up. That's fine, but if your cat rejects a box with a liner because it seems or feels different, you will have much more serious clean-up problems. Similarly, research I described in the section on choosing a litter shows that most cats prefer their litter box to be fragrance-free. Substances that mask odors by adding a smell of their own may drive your cat from the box altogether. They may also create the illusion that you don't need to clean the box when, in fact, you do. A clean litter box that is scooped often doesn't smell.

Litter Scoopers

To keep the litter box clean and fresh, it's important to scoop or sift through the litter one or more times a day (depending on the number of cats you have) to remove any solid waste and clumps. To do this, you need a litter scooper—a long-handled plastic or metal utensil that resembles a slotted spoon or a beach shovel with holes. During the scooping process, the clean litter passes through the slots back into the box, while the clumps of waste remain in the scoop to be discarded.

When choosing a scooper, you want to be sure you buy the right one for the type of litter (traditional clay, clumping, silica gel, corn or wheat granules, pellets) you are using. Some scoops work well with almost any type of litter, while others work best with very fine litters. Make sure the scoop you choose is sturdy. The handle should be comfortable to grip, and the scoop itself shouldn't bend or crack when you're spooning deep under clumps of litter. Beware of adorable-looking scoops shaped like cats and other animals—some of them are very flimsy and break easily.

A few well-known brands include the Booda Jumbo Litter Scoop, VO-Toys Deluxe Cat Litter Scoop, Petmate Jumbo Scoop, Litter Pearls Litter Scoop (designed for pearl litters but works with all types), Smart Cat Scoop & Holder and Handy Stand Scoop (the last two come with holders so you don't have to lay your dirty scoop on the floor).

For less mess, you can also choose a battery-powered scoop. The Pet Crew Power Scoop is a large vibrating scoop that picks up a substantial amount of litter and, with the push of a button, gently shakes the clean litter through the slots, leaving the clumps to be disposed of.

Two other litter box cleaning tools should also be mentioned: Litter Valet and Scoop'n Toss Litter Scoop. Each is a one-piece device with a tubelike, easy-grip handle and a plastic waste bag. Litter is scooped up and sifted, and the waste is funneled down the tube and into the bag in one simple step without your having to come in contact with anything messy.

Another utensil for scooping that has become popular with cat owners is a stainless steel or Teflon-coated spatula. You can

An assortment of litter scoopers

buy these in cooking or kitchen appliance stores. If you plan to use traditional clay litter, in addition to any of the scoopers I've just described, you will also need a long-handled spoon or spatula (without slots) to remove accumulations of wet litter that have not formed into clumps.

Litter Tracking Mats

Each time your cat steps out of his box, litter will cling to his paws (especially the fine granules of the clumping litters) and can subsequently be tracked throughout the house. Eventually, litter can turn up everywhere—on carpets and furniture, on your shoes and clothes and even in your bed. Litter tracking mats help lessen this problem by trapping the litter that sticks between the cat's paws into a grid-

Litter tracking mat

like mat that your cat walks across.

The mats are designed to fit close against a litter box. As soon as the cat exits the box and his feet touch the mat, the texture of the mat causes the paws to open and the excess litter falls through the grids where it can be emptied back into the litter box or disposed of.

Popular brands include Litter Welcome Mat, Van Ness Trackless Mat, Omega Paw Cleaning Mat, Booda Clean Step Litter Pan Mat and Litter Buster Mat. Some are available in regular and extra-large sizes.

Litter Box Liners

Litter box liners are made of heavy-duty plastic and are designed to help stop spillage when you're cleaning out a litter box. The plastic liners come in various sizes to fit different styles of litter boxes, and some come with drawstrings. Instead of pouring litter directly into the litter box, you line the bottom with the plastic liner and then add clean litter. When the box needs cleaning, you pull up the drawstrings or pick up the edges of the liner and close them to form a bag that can be lifted out of the litter box and disposed of. Plastic liners work best with rimmed litter boxes because the top rim helps keep the liner in place.

Liners make clean-up easier and help to preserve the life of the litter box by keeping urine that sometimes pools in the bottom from soaking into the tray to cause odors. On the negative side, plastic liners can exacerbate litter box odor if the cat's urine accumulates in the gathers and folds. And if your cat has sharp claws and digs to the bottom of the box as if he's searching for buried treasure, he'll probably tear or puncture the lining. And urine that seeps through tears or punctures will collect in the bottom of the box and become very smelly. This can also be extremely messy when you take up the plastic liner to dispose of it because soiled litter will be falling out of the tears in the bottom. If your cat does this, forget about liners completely.

You can also buy sifting litter liners. These, too, are plastic and are perforated. They work best with clumping litter. You simply stack several layers of the perforated liners in your litter box and then pour clumping litter on top. When it's time to clean the box, you pull up the top liner. The clean litter will sift back through the holes into the box, while the soiled clumps stay inside the liner to be disposed of. This process is repeated again and again.

Litter Box Deodorizers

Deodorizers, in liquid, powder and granule form, are sprinkled or sprayed on litter. Their purpose is to be effective enough to remove the offensive odor of cat urine and feces, but not so overwhelming that they discourage cats from using the litter box. Generally, deodorizers in liquid form are sprayed onto the litter, while those in granule or power form are spread in a thin layer around the bottom of the box before you add traditional clay litter, or sprinkled on top of the litter after you add the clumping variety.

Deodorizers with a strong scent, like flowery-scented litters, are often marketed to appeal to cat owners, but they do not appeal to cats. Ordinarily, if you scoop the litter box daily, clean the box and change the litter regularly, you shouldn't need to use a deodorizer to control odors. However, some cat owners like to use deodorizers as an extra safeguard.

The cat's sense of smell is extraordinary, and it starts at birth. The presence of pheromones in cat urine plays an important role in toileting habits. When a cat eliminates, he typically enters the litter box and *sniffs* and then he moves around in a circle and scratches the litter to make a shallow depression and eliminates. When he's finished, he *smells the soiled area* and scratches again to cover up his waste. The scent that remains draws him back to his box. If you attempt to cover up a cat's scent with deodorizers that reek of perfumes or chemical smells, he may develop an aversion to the litter and decide to urinate and defecate anywhere else but in his litter box.

I don't want to imply here that you shouldn't use litter deodorizers. Rather, if you do use one, pick a brand that is unscented or contains a *mild*, pleasant fragrance, not one that smells like a rose garden. Some popular brands include Arm & Hammer Super Scoop Cat Litter Deodorizer (with baking soda), Lambert Kay Fresh 'n Clean Cat Litter Deodorizer, Simple Solution Cat Litter Odor Eliminator, Nature's Miracle Litter Treatment, Tidy Cats Cat Deodorizer, Calgon Purrfectly Fresh and Odor Trap.

Soiled Litter Disposal System

LitterLocker is an innovative disposal system for soiled litter. A few years ago, a Canadian baby-monitor manufacturer set out to develop a diaper disposal system that used disposable refills. In the early stages of product development, the idea of using this concept for soiled cat litter emerged, and a team of industrial designers came up with a hygienic, easy-to-use product for cat owners. It's now being marketed in the United States as the Petmate LitterLocker, made by Doskocil Manufacturing Company.

Resembling a disposal unit for baby diapers, the non-skid LitterLocker pail (lined with a seven-layer, odor-proof barrier bag) sits beside the litter box. You just scoop up the soiled litter, drop it in the pail and turn the handle to seal the bag. LitterLocker can store up to two weeks' worth of soiled litter completely odor-free. When the bag is full, you tie a knot, trigger a childproof cutter and dispose of the bag in the trash. Packs of refill bags come on rolls mounted on a spool.

The LitterLocker soiled litter disposal system

4

Feline Bathroom Etiquette

Most of us assume cats will use a litter box because it's their natural tendency, an inbred characteristic. As a result, we don't think a great deal about the kind of litter box and litter we provide and are perplexed and annoyed when our cats reject them and start urinating and/or defecating in inappropriate areas of the house.

By paying attention to some basic guidelines—providing your cat with a litter box and choosing an appealing litter and carefully watching and listening to the subtleties of your cat's communication—you will be well on your way to establishing a problem-free relationship. In chapter 3 we looked at all the different types of litter and boxes you can buy and what kinds of features cat prefer. In this chapter I'll fill you in on what your cat really wants from his litter box experience and give you some guidelines for preventing problems before they start. I'll also explain how to litter box train a kitten—a very simple task!

Cats are easy to litter train because they have a natural desire to cover their urine and feces. They also do not urinate or defecate where they eat, sleep or rear their young. Animal behaviorists say this characteristic is innate and is related to their survival in the wild; they do not want to leave any evidence of their presence that will alarm potential prey or direct a larger predator to their territory. Domestic cats retain this instinct, and even very young kittens will relieve themselves away from the nest as soon as they are able to do so.

Introducing Your Cat to the Litter Box

Litter box etiquette should begin when a new cat is first introduced to your home. Creating good habits in the first few days is easier than trying to correct bad ones later. Cats are fastidious animals and have a natural instinct to eliminate on a soil-type surface where they can bury their waste. This is why using a litter box comes naturally to a cat. In reality, it's not necessary to "litter box train" a cat following the same rigid schedule one would use to housebreak a dog. Indeed, most cats don't need to be taught how to use a litter box.

By the time they are weaned, most kittens have been taught by their mothers to use a litter box. They learn by observing Mom, aided by olfactory clues. Bonnie Beaver, D.V.M., writes in *Feline Behavior: A Guide for Veterinarians* that "kittens have a natural tendency to 'earth rake' loose sand and dirt as a prelude to the use of this behavior in elimination. Around 30 days of age a kitten begins to spend time in a litter box or in soft dirt, moving the particles from one side to another." Normally, adds Beaver, a newly acquired kitten does not have to be litter trained. Some cats, however, including orphans or those who live outdoors, do not have the opportunity to learn, so you must educate them.

If you have a kitten who does not know what the litter box is for, don't be downhearted. Teaching a cat to be clean indoors is very easy because you will be instructing an animal whose natural inclinations are *not* to soil his nest.

Coming Home

Always plan to bring your new kitten or adult cat home on a day when things will be relatively quiet at home. The cat should always travel in a pet carrier, preferably lined with a towel or small blanket for warmth. His litter box, clean litter, scoop and all other supplies should be in place before the cat enters your home. Once you get home, keep the cat in his carrier and let him observe the surroundings and become familiar with the strange faces, noises and smells of his new home. If other pets are in residence, keep them away so they don't frighten him and add to the confusion.

Confinement and Teaching the Basics

Even if there are no other resident pets, never give a new young or adult cat free access to your home right away. He may relieve himself in places that are easily accessible to him but difficult for you to locate. It's better to place a clean litter box in a small room, such as the bathroom, kitchen or bedroom, and confine the cat there until he has time to adjust to his surroundings and learns to use the box reliably. Temporarily, you should also place everything else the cat will need in the confinement room: scratching post, bedding, toys, food and water. Do remember to keep his food and water dishes and bedding as far away from the litter box as possible. (Once the cat is fully litter box trained and allowed free run of your house, you can move the food and water dishes to another location.)

Make sure the confinement room you choose is free of hazards. In the bathroom, for instance, keep the toilet lid closed. In the kitchen, block off the areas behind the refrigerator or other appliances with a screen or corrugated cardboard so the cat can't squeeze behind them. Always check the refrigerator and dishwasher doors before you close them to be sure the cat hasn't crawled inside when you were not paying attention. In the bedroom, be sure all closet doors and drawers are closed, and don't leave any medications or small items such as jewelry on the nightstands.

Take the cat, in his carrier, to his confinement room and close the door. Place the carrier on the floor. Talk lovingly to the cat, and after a short time, open the door and let him come out on his own to stretch his legs and investigate things. Keep the carrier door open so he can dart back inside if he's afraid.

In the case of a kitten, try to add to his bedding something familiar that contains the scent of his mother and/or littermates, such as a towel or a blanket they slept on. For an adult cat, something from his former home or shelter will make him more comfortable during the transition period.

Your first act once you let your cat out of the carrier should be to point out the location of the litter box. If the cat is afraid to come out of the carrier, don't force him. Fill his food and water dishes, leave the room and let him come out to explore his surroundings when he's ready.

Do remember that "confinement" does not mean "prison." Plan to spend lots of time with the kitten or adult cat, speaking to him in a calm and reassuring voice, playing with him and making him feel welcome in his new home. Don't ignore any other cats or dogs in residence; they need plenty of attention and affection, too.

Indoor/Outdoor Toilet

It is a fact that indoor cats live longer lives than outdoor cats. According to the Humane Society of the United States, an average indoor cat's life expectancy is 17 years or more, while that of an outdoor cat is between 2 and 4 years. Cats who remain inside their house or apartment, a place of business or a public building can live long and happy lives without ever venturing outdoors. Outdoor dangers include moving vehicles, environmental hazards, pet-hating neighbors, fleas, vicious dogs, other cats who spread diseases or who are determined to fight over their territories and, in certain parts of the country, coyotes and other wild animals and even large birds of prey.

In spite of these facts, some owners believe their cats are deprived if they aren't able to go outdoors. It is possible to let a cat enjoy the outdoors without subjecting the animal to danger by building a screened-in porch or a cat run in the yard. Otherwise, cats should stay indoors where they will be safe. If your cat is fully immunized, however, and does do his business outside part of the time, *please keep him inside during the night and let him urinate or defecate in his indoor litter box.* Dangers multiply at night.

Remember that this is a very stressful time for the newcomer. In chapter 1 we learned a little about pheromones. They can have a calming effect on cats, especially the ones that come from the cat's cheeks and forehead. When a cat rubs his face against you or an object, he's indicating with facial pheromones his comfort in his surroundings. A synthetic version of these naturally occurring pheromones is available commercially. The product is called Feliway, and it comes in both spray and plug-in diffuser form. Designed to mimic the cat's facial pheromones, Feliway is very effective in calming and comforting a cat in stressful situations. During the transition period, use Feliway in the confinement room, either by plugging the diffuser into an electrical outlet or by spraying it on objects in the room and the doorway (at a height he can smell)—areas the cat might normally rub against to scent mark.

Most cats will eliminate after they awaken, after they eat and drink and after play periods. With kittens, there is a short interval of about 15 to 20 minutes between when they eat, drink or play and when they eliminate. So the first thing every morning, about 15 to 20 minutes after meals and play periods, and before going to sleep at night, begin to reinforce litter training by taking your cat to the box and placing him inside. Chances are he'll know what to do and has already eliminated in his box. Be sure to praise him lavishly (but not loudly) when he does. Let him know that what he has done pleases you tremendously. You don't need to use the same word or phrase each time; your tone of voice will convey your affection and enthusiasm. Express your pleasure with your touch, too, by stroking him. Your hands should always communicate affection. And each time you express your approval you will be positively reinforcing the behavior you praise.

If the kitten doesn't understand what the litter box is for, put him gently inside and try to stimulate his interest by ruffling up a little litter with your finger. Be sure the litter is clean and that you wash your hands afterwards. Don't hold the kitten or cat rigid in the box; let him jump in and out at will. For the first few days, leave a *tiny* bit of urine and feces in the box to entice your cat to go there the next time.

If the cat urinates or defecates outside the box, pick up as much as you can and deposit it in the box. A whiff of his own waste should stimulate his natural instincts to eliminate there. If you find an accident, clean it up promptly and *don't* yell at or strike the cat. Shouting and physical punishment cause fear and anxiety and actually inhibit learning. Your cat will consider such behavior to be a negative experience with the litter box and may avoid it entirely.

After Confinement

It may take a week or two of confinement before your cat uses his litter box properly every time. Once the cat is familiar with his litter box, you can start allowing him to roam freely through the house. For now, *don't move the box* because you'll only confuse the cat.

Accidents occasionally do happen with a new kitten or adult cat. If an accident occurs away from the litter box, *do not spank or rub the cat's nose in the mistake,* especially if you didn't catch the cat in the act. Animals don't relate past actions to current consequences, and your cat will not understand discipline for something that happened previously. So many young animals are intimidated this way by ill-advised owners, and punishment after the fact may only make the cat repeat his wrongdoing.

If you catch your cat in the act of urinating or defecating away from the litter box, pick him up, say "no," carry him to the box and deposit him inside. You must clean the accident area thoroughly and remove all traces of stains and odors to prevent your cat from soiling the same area again and again. Even if you can't smell anything, your cat still can, so follow the cleaning suggestions in chapter 7.

Where Should the Litter Box Be Located?

Cats prefer seclusion, especially when they are eliminating—a position in which they feel very vulnerable. So it's very important from day one to find the appropriate private, low-traffic but accessible area in which to locate the litter box. Here are some suggestions to help you determine the best place.

- *Remember, cats do not like to eliminate close to where they eat or sleep.* The most important guideline is never to locate the litter box near your cat's food or water dishes or his bed. You're only asking for problems if you do—major problems!

- Remember, too, that cats are territorial by nature. In the wild, cats select and establish their territories in proportion to the number of other cats in a given area. When there are two or more cats in a home, there can be dramatic confrontations among individuals as each delineates his territory. For that reason, you should place the litter boxes in different locations around your home, well away from one another. Doing so will let adversaries avoid one another, if they wish. Do not place the litter boxes near one another in the same area.

- Every box must be easy for the cat to get to. This is especially important for very old and very young cats, but it applies to all cats.

- If there is a dog in residence, put the box in a place where your cat can eliminate without being annoyed. Keep in mind, too, that some dogs enjoy eating cat feces. The solution may be to place the box up off the floor, to choose a covered litter box or to place the box behind an opening that is large enough for the cat to fit through comfortably but is too small for the dog.

- You can hide any litter box behind a privacy screen. These are 30-by-19-inch trifold screens made from tough, water-repellent corrugated plastic, covered with stain-resistant fabric. Another way to provide privacy is to conceal the litter box inside a carpeted bench that your cat can sit or play on top of. Ask your pet supplies dealer about these, or check the pet supplies mail order catalogs listed in the appendix. Remember, though, that your cat must still be able to get to the box easily.

- High-traffic rooms, such as the kitchen, family room and den and other places where there is a great deal of noise, are not good choices for litter boxes.

- Basements and garages are bad locations because they are remote and dark and generally have cold cement floors. These are places cats do not want to go. Basements and garages often

contain odds and ends that can scare a cat and drive him away
from his litter box. A kitten, or especially an older cat, may
not have enough bladder control to reach the box in time. It's
also very inconvenient to clean a litter box that's placed in
such an out-of-the-way spot.

- The bathroom can be an excellent location. It's easy and
  convenient for your cat to locate and easy to clean. Just
  don't choose an area that makes it difficult for the cat to
  complete the elimination sequence, such as wedging the
  box under the sink (where he doesn't have enough room to
  turn around comfortably) or putting it in the bathtub. You
  must also make sure the bathroom door stays open at all
  times!

- The utility room is another good choice, particularly if noise
  doesn't spook your cat. It's cozy and more pleasant than the
  basement. Depending on what's in the utility room—clothes
  washer, dryer, air conditioner or furnace—the only disadvan-
  tage may be the sounds of a large appliance in operation. A
  cat can become so frightened by the noise of the spin cycle
  kicking in that he will subsequently refuse to use the box.

- A low-traffic, quiet area such as a spare bedroom is another
  good choice. Again, the door must remain open at all times.
  You could install a cat door into the room to protect the cat
  from being bothered by children or other pets.

- Always place the litter box or boxes where they can be easily
  accessed by all cats in your household and away from carpeted surfaces or walls that are difficult to clean if the cats spray.
  If you must use a carpeted area, protect it by putting a washable mat underneath.

- Once you have determined the right location for the box,
  keep it there. Moving it will only confuse your cat, and he
  might decide to do his duty where the box used to be or to
  find another secluded spot, such as on the carpet behind the
  furniture or in a potted plant.

Cats, Litter Boxes and Allergies

Cats, unfortunately, do produce allergic reactions in humans, and sometimes they can be severe. The major cat allergen, Fel d1, is a protein that has been identified in the sebaceous glands of the skin (called "dander"), as well as in cat saliva. Keep in mind that cats spend half their waking hours licking themselves clean. After Fel d1 dries on the cat's skin and hair, the tiny particles flake off, become airborne and circulate throughout the house, where they can be inhaled with every breath.

It is thought that feline urinary proteins may also be allergenic; therefore, when cats live indoors with allergy sufferers, their litter box should be scooped and cleaned frequently, and it should *not* be located near ducts that circulate heat or air throughout the house or near areas regularly used by allergy sufferers, especially their main activity rooms and bedrooms.

Equally important, allergic persons should *not* clean cat litter boxes or should wear a mask if they have to do so. Wiping the cat's coat (paying particular attention to areas the cat licks most often) once a week with Allerpet/C will help to remove cat-related allergens before they have a chance to enter the allergy sufferer's environment.

How Many Boxes Do You Need?

How many litter boxes you need depends on the number and temperament of cats in your home and the size of your house. Many cats don't mind sharing their litter boxes, but some do. Some cats may be attracted to the scents left by their housemates, while other, more sensitive cats may avoid using the box entirely. Feline behaviorists generally advise having one more litter box than the number of cats in your home. In other words, one cat should have two boxes, two cats should have three, three cats four, and so on.

Don't make it difficult for your cat to reach the litter box, or problems may develop. If your house is large and/or has more than one story, you should place an additional litter box at opposite ends of the house or one on each floor. Kittens and older cats also have a harder time controlling their bladder and bowels and may need more boxes.

15 Ways to Stop Litter Box Problems Before They Start

1. Provide one litter box for each cat in your house, plus one extra. Add more boxes for each level in your house.
2. Place litter boxes in quiet, private, easily accessible and well-ventilated areas. Avoid dark, damp basements and garages, distant bedrooms, congested household areas, washing machines, television sets, furnaces and any areas where your cat could be disturbed by small children, other animals and sudden loud sounds. The litter box should be placed in an easy-to-reach area that you can scoop every day. Never place it in a remote area that you forget about until it starts smelling bad. Remember, out of sight, out of mind.
3. Remember that cats are territorial. In a multiple-cat household, space litter boxes far enough apart to avoid territorial confrontations.
4. Make sure the litter box is big enough for your cat to stand up and turn around in, without any parts (especially the rear parts) hanging over the sides. If you use a covered box, make sure the cover is high enough that your cat can stand up straight and not bump his head. Some cats will not use a covered box; if your cat seems reluctant to enter one, remove the cover and forget about it.
5. Scoop the box at least once a day, and keep it clean and odor-free.

6. Do not place the litter box near the cat's food and water bowls or sleeping area.

7. Do not move the litter box around your house. Always keep it in the same location.

8. If you are introducing a new kitten or cat to your home, use the same type of litter box and litter that he used in his previous home.

9. Keep the box filled with the amount of litter your cat prefers. You may need to experiment with different amounts of litter until you decide what's right for your cat. Finicky cats sometimes object to too much or too little substrate.

10. Do not use heavily scented litters. They do not mask smells very efficiently, and cats are often repulsed by their odors.

11. Do not use heavily scented litter deodorizers. Choose a brand that is unscented or contains a mild, pleasant fragrance. Regular baking soda or Arm & Hammer Super Scoop Cat Litter Deodorizer with baking soda helps counteract odors and soak up moisture and can be used with any litter. Spread a thin layer around the bottom of the box before adding the litter. And occasionally sprinkle a little more on top after scooping and cleaning.

12. Do not place room deodorizers or scented air fresheners near the litter box.

13. Try not to force a new brand of litter on your cat. Doing so often causes cats to stop using their litter boxes and start looking for other areas in which to urinate and defecate.

14. Plastic litter box liners can make changing the litter easier, especially if your cat does not scratch up a great deal of litter when he buries his waste. Use a liner, by all means, if it makes you change the box more often, but if your kitten or cat avoids the box, then dispense with the liner.

15. If you do choose to use a liner, attach it securely to the litter box so the cat can't pull it off as he turns around, scratches, digs and completes the elimination sequence, or use a rimmed litter box.

Keep Your Cat's Litter Box Clean

Just as people abhor using dirty rest rooms, cats dislike foul-smelling and damp litter boxes. The main reason cats refuse to use their litter boxes is that they smell bad. The best way to control odors, no matter what type of litter you use, is to keep the box clean and dry.

It's pointless to try to mask any odors because cats, with their highly developed sense of smell, will never be fooled. Dr. Nicholas Dodman, professor of Behavioral Pharmacology at Tufts University School of Veterinary Medicine and director of the Behavior Clinic says, "The cat's nose is at least 100 times more sensitive than the human nose." So if the litter box smells even a little bit bad to you, it's *definitely* more offensive to your cat. Here are some tips that will help you keep your cat's toilet clean and fresh:

20 Ways to Control Litter Box Odor

1. Provide at least one litter box per cat. Place them in well-ventilated areas.
2. The most effective step you can take to combat odor is to remove urine and feces at least once a day if you have one cat and two to three times a day if you have several cats. While the smell of urine may be alluring in low concentrations and attract a cat back to the litter box, it will be repulsive in high concentrations in a messy litter box.
3. Place the box on a floor that's easy to clean, not on a carpet. Put a large terrycloth towel, a washable mat or a disposable pad under each box to absorb any "mistakes" (some cats hang their rear ends out the side of the box) and to reduce tracking each time the cat steps out of his box.
4. Maintain about a two-inch layer of litter in each box. The amount of litter you use plays an important role in odor control. Using too little can make the box smell foul because the urine pools in the bottom and there is not enough substrate to satisfactorily absorb it. Using too much can also makes the box smell; urine will pool on the bottom while the top layer of litter looks clean.

5. Do not use regular shredded newspaper in the box; the ink comes off on the cat's feet and the paper quickly becomes smelly when saturated with urine. This might also inspire a cat to relieve himself on any newspaper in the house!

6. Clumping, pelleted and silica gel litters are the easiest to maintain. Clumping litters tend to stay dry because the urine forms into hardened clumps before it can break down into ammonia. When the clumps are sifted regularly, the box stays clean much longer than with non-scoopable clay litter. Use a sturdy slotted scoop to remove soiled urine clumps and dried feces, and remember to scoop all the way down to the bottom of the box.

7. Clumping litter always lasts longer and smells fresher when very small fragments are removed. Try not to shatter the clumps as you remove them. If they do fragment too easily, try a brand that clumps harder and faster.

8. Non-scoopable clay litter is more difficult to keep clean. Saturated litter that collects at the bottom of the box can become disgustingly smelly. First, use a sturdy slotted scoop to lift out any dried feces. Next, use a long-handled plain (not slotted) scoop or metal spatula to remove any mounds of wet litter. Try not to stir the wet litter around too much; you'll simply distribute it throughout the box and intensify the smell.

9. When you have finished scooping, always add some fresh litter to replace the soiled litter that has been removed.

10. If you're using a traditional clay litter, cover the bottom of the box with baking soda or Arm & Hammer Cat Litter Deodorizer (which contains baking soda) before refilling it with fresh litter. With scoopable litter, pour the baking soda deodorizer on top of the fresh litter and mix it in. Remember to wash your hands thoroughly when you have finished.

11. Dispose of the soiled litter as soon as possible. Be sure to read the manufacturer's directions for disposal. Some scoopable litters can be flushed down the toilet (although large amounts may clog the toilet), while others, especially the heavy-duty formulas, are *not* flushable. Traditional clay litter should never be flushed down the toilet. *Do not flush any kind of litter if you have a septic system.*

12. Here's one efficient disposal solution: Line a small, covered plastic container with a sturdy drawstring or handle-tie trash bag. Place the container near the litter box. Deposit all scooped clumps or mounds of wet litter into the container and snap the lid securely shut. At the end of the day, tie up the bag and drop it into an outside trash container. You might also want to check out the new smell-free LitterLocker (described in chapter 3), a hygienic device designed to dispose of cat waste.

13. Diet plays a significant role in litter box odor. Although cats require a lot of protein in their diet, diets that contain too much meat protein consistently produce intense, foul-smelling feces. Generally, there is less smell when the diet is better balanced. Certain diseases and intestinal infections can also worsen the odor of feces. The amount of food also influences how much stool cats defecate. Some foods can minimize the volume of feces, which results in a lot less odor. Consult your veterinarian for the appropriate diet.

14. How often you empty the box and change the litter depends on the size and number of cats you own, the cats' habits, the style and number of litter boxes in your household, the type of litter you use and how often you scoop. For a single cat, generally you should change non-clumping litter once a week and most clumping, pelleted and silica gel litters once a month. However, absolutely do change the litter whenever you detect odors or when the litter is soggy or full of clumps.

15. When you empty the litter box, scrub it (plus all scoops and spoons) thoroughly with hot water and mild dishwashing liquid or white vinegar. If your litter box has a cover, don't forget to clean it regularly, too. Avoid using harsh household disinfectants or cleaning products; some can be toxic to cats. Those containing ammonia will heighten the odor of urine. Others may leave a perfume residue that can deter a cat from using the box. Rinse and dry the box thoroughly.

16. Air dry the box so you are sure it is thoroughly dry. If you can, air dry it in the sun because direct sunlight is nature's own germ killer and air freshener. If you live in an apartment, litter boxes can be air dried in the sun on the windowsill of an open window (screened, please, for your cat's safety!).

17. When you clean the litter box, remember to launder the mat under the box and to mop the floor around the box with a deodorizing cleaner.

18. Once a month if you use regular clay litter, or every other cleaning if you use clumping litter, use chlorine bleach diluted with water to disinfect each box (about one-quarter cup mixed in a gallon of warm water). Do go easy on the bleach; occasionally a cat will stop using his litter box because the odor of the bleach is too strong. Keep rinsing until all traces of bleach scent are removed. Dry the box thoroughly before refilling it with fresh litter.

19. Replace plastic litter boxes from time to time because they eventually become impregnated with urine odor. If the box doesn't smell clean even after you have washed it, throw it away and get a new one.

20. For odor problems in other parts of the house, you need products formulated specifically to neutralize urine and feces odors on floors and carpets and not merely a deodorant. Read more about these in chapter 7.

5

Specific Solutions for Specific Problems

Inappropriate elimination is the number one reason owners surrender their cats to animal shelters. It can be impossible to live with the smell and mess. It may seem the cat is being spiteful or stubborn, but in reality, your cat is behaving in a normal feline way. The problem is, what works for him just doesn't work for you.

As I mentioned in chapter 2, when a cat suddenly starts to soil the house inappropriately or to spray urine, the first step is to take the cat to the veterinarian for an examination to rule out medical causes. A variety of medical problems that can cause a change in litter box habits are discussed in that chapter. Once your veterinarian rules out any medical reason for house-soiling or urine spraying, an environmental or behavior problem is likely. Cats, quite often, will urinate outside their litter boxes or spray urine on vertical objects to call attention to a problem.

To resolve litter box issues and create a closer relationship, you need to understand why your cat has suddenly abandoned the litter box. Because once you understand the problem you can implement the solution.

Spraying or Squatting: What's the Difference?

It is important to determine whether a cat is house-soiling or urine spraying because they are different problems with different solutions. House-soiling is when a cat squats and lowers his hindquarters to urinate or defecate on horizontal surfaces outside the litter

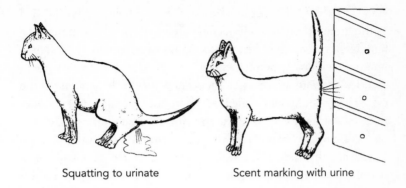

Squatting to urinate   Scent marking with urine

box, such as on carpets or floors. Urine spraying, a form of territorial marking, is when a cat stands erect, facing away from the object to be sprayed, holds up his tail with the tip twitching and forcefully directs a stream of urine toward vertical surfaces, such as walls, chairs and furniture. Some cats also make treading motions with the hind feet during the procedure. The amount of urine deposited is not always the same, but it is always visible on the sprayed surface. Owners may not always observe their cats in the act inside their homes, but, usually, they will smell a strong urine odor or find urine on vertical surfaces.

Urine Spraying

As I discussed in chapter 1, scent marking is a form of scent communication for cats. Marking can take many forms, some of them affectionate and others frustrating for cat owners. Cats can mark by rubbing a person or an object in the household with their chins, cheeks or foreheads—areas that contain scent glands. Cats may also mark by spraying their urine. A fatty ingredient in the urine leaves a pungent odor that clings to objects and leaves a strong smell. There are also small anal glands on either side of the rectum that release a liquid to mark the cat's feces with a specific identifying scent.

Pam Johnson-Bennett, feline behaviorist and author of *Think Like a Cat*, advises owners to "draw an imaginary line that divides your cat in half. The scent glands on the front half could be labeled as the 'friendly' pheromones. These are used when a cat is marking familiar territory that he considers the heart of his nest. These pheromones have a calming effect . . . that reflects the cat's sense of security." Beware, however, of the scent glands at the rear end of the cat. These could be labeled as "alarming" because pheromones released during the act of urine spraying are related to stress.

It is male cats who primarily spray urine, although the practice is occasionally seen in females. When unneutered male cats reach maturity, around eight to nine months of age, they become possessive about their home territory, indoors or out, and begin to spray urine to establish and define it. The odor of their own urine may make them more comfortable and self-confident. They spray on vertical surfaces so that the resulting scent marks at a level that other cats can easily smell.

Urine spraying, like house-soiling, may also be triggered when a cat feels stressed or displaced, such as when a new pet, baby, spouse or roommate comes into the household; when there are changes in the environment; when the litter box is dirty or is used by too many cats; when cats are having social disputes; when a cat can smell or see another cat outside the house; and even when a cat is not getting enough attention. Emotional problems are common in multi-cat households—in fact, the more cats in your household, the greater the likelihood of spraying.

The Solution to Urine Spraying

Once medical causes are ruled out, the single most effective treatment for suppressing the obnoxious habit of urine spraying by both males and females is castration or spaying. You may be familiar with the term "neutering," which also means spaying or castration. Ninety percent of male cats and 95 percent of females stop urine spraying within three weeks to three months after being neutered. And when the procedure is performed before sexual maturity, the vast majority of cats never begin to spray.

Spaying and castration are terms for the surgery performed by a veterinarian to remove the reproductive organs. Female cats are spayed—that is, their ovaries and uterus are removed. Male cats are castrated—that is, both testicles are removed. Spaying and castration are irreversible procedures.

Spaying or castrating does not adversely affect a cat's personality, but the operations do make cats of both sexes less noisy and aggressive, more affectionate, playful and content. Males, if allowed outdoors, are less likely to roam and be drawn into fights over females in heat. Neutered cats of both sexes usually live longer and healthier lives. Spaying eliminates the possibilities of uterine and ovarian diseases and greatly reduces breast tumors and mammary cancer in females.

For the small percentage of cats who continue to spray, follow-up treatment consists of several steps:

- Clean urine marks daily with an enzymatic cleaner.
- Frequently scoop and clean the litter box.
- Close drapes or block windows to prevent the spraying cat from seeing or interacting with outdoor cats.
- If the cat is only spraying one or two areas in the house, feed at those spots during regular mealtimes.
- Consider anti-anxiety or mood-stabilizing medications such as fluoxetine (Prozac), buspirone (Buspar), paroxetine (Paxil), alprazolam (Xanax), amitriptyline (Elavil) and others. However, drugs should *only* be used under veterinary supervision and as part of a comprehensive behavior modification plan. Your veterinarian can prescribe a drug therapy that is safe because these medications can cause serious side effects if misused.

Feliway: A Synthetic Pheromone

Another solution for the small percentage of cats who don't stop spraying is behavior modification using Feliway, a spray that contains analogues of natural feline facial pheromones. Feliway can be very calming to cats; it reduces their urge to spray and mark by reproducing the familiar smells normally produced by a cat when she rubs against objects or people and deposits her own facial pheromones.

When Feliway is sprayed onto objects in the cat's environment, it counteracts the cat's alarm pheromones with calming ones and stops urine marking. Feliway comes in both spray and plug-in diffuser forms and is available at most pet supply stores. It must be sprayed daily for 30 days onto spots where the cat has sprayed and on tempting vertical surfaces. If you're too busy to hand-spray, just use the plug-in diffuser in the room where marking has been a problem. If the cat has marked several rooms, depending on the size of your house, you may need to place an additional diffuser in each room.

Originally, Feliway was created to treat urine marking behavior, but it is also effective for various other stress-related problems that are outlined later in this chapter and in chapter 6.

House-soiling

In addition to the medical problems mentioned in chapter 2, there are many environmental and behavioral causes of house-soiling. Odors, a dirty litter box, unacceptable litter, an inaccessible box, a covered box, a box that is too small, a box located in an area a cat does not like or is frightened of, a dislike of plastic litter liners, a lack of privacy or a negative experience associated with the box may all cause a cat to avoid eliminating in the litter box. There may be other triggers: tension or aggression between cats in the household; a new cat, dog or baby in the home; a new cat in the neighborhood who is visible from a window; long periods of separation from family members; the loss of a beloved companion; home renovations; and even changes in a cat's diet.

Treating all these house-soiling problems consists of three steps:

1. Identify and remove the cause of the problem. If the cause is medical, for instance, treat the disease or condition. If the cause is environmental or behavioral, determine why the cat is disturbed and either change the environment or minimize the stress factors.

2. Redirect the cat to use the litter box; for instance, by keeping the litter box scrupulously clean, scooping and emptying it more often, providing a litter the cat likes, removing a plastic litter box liner or providing more privacy to make the box as attractive and accessible as possible.

3. Prevent the cat from returning to previously soiled inappropriate areas by using an enzymatic cleaner to thoroughly clean and eliminate the odors and stains. After eliminating on an inappropriate surface for a time, a cat often develops a preference for that particular surface and wants to continue to urinate or defecate there—even when other issues have been resolved. It's also very important to make previously soiled spots unusable, for example, by placing deterrents at the site. This may include covering the affected area with barriers such as heavy plastic or aluminum foil. You can also try to temporarily make the favored spot into a feeding or sleeping area by placing food dishes or bedding there.

Finicky Breeds

It is interesting to note that according the Morris Animal Foundation, Persians and Himalayans are more prone to elimination problems than any other breeds. Seventy-five percent of these cats decide at some time in their lives that they don't like their litter boxes.

What About Punishment?

Punishment does not work! Punishing a cat for house-soiling or urine spraying (especially when the cat is near the litter box) will not solve the problem. In fact, punishment will only aggravate matters because a negative experience associated with the litter box will only make the cat more likely to avoid the box. If you catch your cat in the act, calmly pick her up—no rubbing the cat's nose in the "mistake," yelling, slapping or physical punishment— say "no" in a matter-of-fact tone and put her in the litter box.

You can also use other kinds of distractions *if you catch your cat in the act.* (If you find an "accident" after the fact, all you can do is follow the three steps above.) Keep a water pistol or a spray bottle filled with plain water or a small, soft object (such as a cat toy) in a hidden spot near the areas where the cat is eliminating inappropriately. If you see your cat urinating or spraying there, quickly distract her by squirting her with the water gun or spray bottle (setting the nozzle to "stream," not "mist" will be more effective) or by tossing a soft object *toward* her, not *at* her. You can even clap your hands loudly or stomp your foot. Don't use your voice, and be careful *not* to scare the cat. You just want to interrupt the behavior.

When you use these kinds of remote distractions, your cat will learn to associate the consequences with her behavior, not with you, according to Wayne Hunthausen, D.V.M., a feline behaviorist. "It is important that the cat not associate punishment with the owner," adds Hunthausen, "or the bond between the pet and owner will quickly deteriorate." It's also beneficial to reward your cat with praise and a favorite treat when she uses the litter box correctly.

The best solution is environmental or behavioral modification to make the litter box as attractive and accessible as possible, while making other areas less appealing and unsuitable for urination. It is also very important to clean any soiled areas with an enzymatic cleaner to eliminate any odors left from previous urination or defecation. Read more about removing odors and stains in chapter 7.

Problems With the Litter Box

Dirty Box

The number one reason cats refuse to use their litter box is that it smells bad. Just as people dislike using dirty rest rooms, cats dislike foul-smelling, damp, messy litter boxes. If the litter box smells bad to you, it's definitely more offensive to your cat since her sense of smell is at least 100 times more sensitive than yours. The best way to control odors, no matter what type of litter you use, is to keep the box clean, dry and smelling fresh.

There is no substitute for scooping and scrubbing. That is just a fact of life for cat owners. See chapter 4 for more tips on keeping the litter box clean.

Hooded or Covered Box

A covered box needs to be cleaned as often as an open box—sometimes more often, especially if several cats are using the same box. Although some covered boxes come with built-in activated charcoal filter systems, those only filter the air on the outside of the box; inside, odors are trapped under the cover. The litter may also become soggy because of poor air circulation.

If you don't clean the box as often as you should, your cat, with her well-developed sense of smell, has to tolerate these intense, offensive odors each time she goes inside the box to eliminate. It creates a distinct outhouse effect that may be unendurable.

Keep in mind, too, that using a covered box may force a large cat to enter a cramped area where she doesn't have adequate room to perform the elimination sequence of scratching, squatting and turning around. If she feels cramped and uncomfortable in the box, she won't use it. Try getting a bigger box with a bigger, higher cover. If that doesn't work, take the cover off and throw it away.

Undesirable or Inaccessible Location

Where the litter box is placed often causes a cat to avoid it. Cats do not like their litter boxes to be in busy places or areas that are difficult for them to reach. They prefer seclusion, especially when they perform the elimination sequence, so it's very important to find the appropriate private, low-traffic but accessible area in which to locate the litter box. It must be a place where the cat will not be disturbed by people or other pets, where loud noises (such as the washing machine) will not suddenly startle her, and a spot she can reach comfortably and without effort. It should not be someplace that is cold and damp (such as the garage or basement), where a cat would not choose to go.

Many companies sell litter box cabinets that hide the box inside a cupboard or a piece of feline furniture. If you opt for one of these, be sure your cat has no trouble getting through the opening and into the box. Also be aware that in a closed space, a box will smell worse and will need to be cleaned more. And remember that some cats simply will not accept this arrangement. For many suggestions about litter box location, see chapter 4.

Box Is Not Private Enough

When a cat uses a litter box, she wants privacy. Think about it: You wouldn't like going to the bathroom in the center of a busy room while other family members walk by at any time, would you? Of course not! You count upon privacy. So does your cat. She feels very vulnerable while she's urinating or defecating, and she doesn't want to be pestered by children, adults or other animals. If she is disturbed, she'll look for another, more tranquil place to urinate or defecate—hopefully not on your treasured Oriental rug. Find the appropriate private, low-traffic but accessible area in which to locate her box. A box with a hood or a privacy screen may be helpful. Some cats adore them; others don't. You might try placing two boxes in different locations and see which place she prefers.

Box Is Too Close to Eating or Sleeping Areas

Cats are extremely clean and fastidious animals and refuse to eliminate in areas where they eat or sleep. Because their urine and feces are so pungent, cats eliminate away from their nests in the wild and cover their waste to avoid attracting predators back to their home base. Domestic kittens and adult cats retain this inherited instinct and want to relieve themselves as far away from their "nests" as they can. *Always* locate your litter box away from your cat's food and water dishes, bedding and favorite resting and napping spots.

Unacceptable Litter

One important reason cats refuse to use their litter boxes is the litter itself. Cats can have a preference for one type of litter over another, and, indeed, they can dislike certain textures or smells. Tests have conclusively shown that cats prefer soft, sandlike litter, such as the fine-grained clumping type. The next most preferred are traditional clay litters, followed by the alternative litters, such as recycled material, silica gel, wood, plants and plant by-products.

Cats hate change! I've said it over and over in this book, but it bears repeating. Never try to abruptly force a new litter on your cat. Always use the same type of litter your cat used successfully before you acquired her. If, for any reason, you must switch, do it by adding a tiny amount of the new litter in with the old. Once your cat uses that mixture, gradually add a little more of the new litter each day and reduce the old, until the box contains only the new litter. If the cat refuses to use the box, go back to the mixture she will use and start again, only each time mix less of the new litter in with the old. It may take several weeks to accomplish this changeover—and you may never be successful. If your cat insists on a type of litter you do not like, *you must go with your cat's preference.*

You can experiment a bit to see which litter type your cat likes best. Buy a small amount of each type—clumping, non-clumping and the alternatives—put each into a separate litter box and see which one your cat prefers.

Don't use heavily perfumed litters. More often, litters with flowery scents or added deodorizers are marketed to appeal to cat owners and not to cats. Besides, they do not mask smells very efficiently, and cats are often repulsed by their odors. Cats want to be able to smell a bit of their own scent in the litter, not overpowering perfume.

Change in Box Style

A change in the type and style of litter box can disturb your cat so much that she will look for other places to eliminate. Do not abruptly change to a new box because it was on sale or it looks appealing to you.

However, there are times when a litter box needs to be replaced, especially if it's old and smelly. If you must change boxes, keep the old one on hand for a short time. Fill the new box with the litter the cat prefers, place it alongside the old one and let the cat select one or use both. Then, after a week or so, when the cat is regularly using the new box, take away the old one.

Change in Location of Box

Most cats associate the acts of urinating and defecating with a specific location—hopefully, the litter box. If you have to move the box from one location to another, do it gradually over a period of several days, by moving it a foot or so at a time toward the new spot—even though doing so might be an annoyance to you. If your house has more than one story, keep the litter box on the same story.

Box Is Too Small

Is your cat's litter box so small that she can't turn around comfortably or her rear end hangs over the side and she misses the litter when she eliminates? Or perhaps your cat has outgrown her first "kitten" box and needs more room. Litter box size matters a great deal. Pick a litter box that will be roomy enough for your cat to urinate and defecate in several areas and still have some clean spots to stand on. As a general guideline, feline behaviorist Pam Johnson-Bennett advises that "the length of the box should be double the length of an adult cat and the width of the box should be the cat's approximate length. If you have more than one cat and they vary greatly in size, base your measurement on the largest cat." The bottom of the box should be deep enough to give your cat enough "burying" room without spilling litter outside.

A cat who scratches around in a covered box that is too small for her may accidentally dislodge the cover—scaring herself in the process. When that happens, throw the cover away or your cat will forever associate the litter box with a fearful experience.

Home Is Too Big for One Box

When nature calls (especially if there are medical problems) and the only litter box is at the opposite end of the house, or up or down the stairs, your cat may not get there in time. If your house is large and/or has more than one story, place an additional litter box at a different end of the house or on each floor.

Lingering Scent of Cleaning Solution

It is very important to clean each litter box (plus all scoops and spoons) once a week with hot water and mild dishwashing liquid or a little white vinegar and to rinse it thoroughly. You must, however, avoid using strong-smelling household disinfectants or cleaning products. Some can be toxic to cats. Those containing ammonia will heighten the odor of urine. Others may leave a strong residue (even if you can't smell it, your cat can) that can deter a cat from using the box. Dry the box thoroughly before adding fresh litter. Air dry it outside or near an open window, if you can. Don't forget to periodically replace a plastic litter box because they eventually become impregnated with urine and begin to smell even after regular cleaning.

Plastic Litter Box Liners

Some kittens or adult cats often refuse to use a litter box that is lined with plastic, while others don't mind at all. Plastic liners can make changing the litter easier, especially if a cat does not scratch up a great deal of litter when she attempts to bury her waste. However, during the process of digging in the litter and scratching to cover the waste afterward, your cat's tiny claws may catch on the liner and puncture the plastic. Urine can seep through the punctures, collect on the bottom of the litter box and begin to smell foul. Other cats are so determined to bury their waste that they pull up all four sides of the liner into the box to completely conceal the litter. Then, if you haven't straightened things out, they will eliminate on top of the plastic, making cleaning very difficult. And some cats simply do not like the feel of the liner under their feet or the rustle of the plastic as they dig and scratch. If your cat avoids a box with a plastic liner, dispense with it.

Negative Experience Associated With the Litter Box

Punishment, trimming the nails, combing matted fur, dropping medication into the ears, giving the cat a pill and other negative experiences that happen in the vicinity of the litter box can frighten a cat and cause her to look elsewhere to eliminate. A cat with medical problems may also associate her litter box with the cause of pain when she urinates or defecates. Rule out medical causes first. Then *stop* performing any unpleasant experience near the litter box.

Cats may also have a negative experience inside the box if something happens nearby that startles them. The spin cycle on the washing machine may kick in, a shelf may fall down, another cat may knock a vase off the table or something else that seems (to you) to be unrelated could scare her. But because it happened when your cat was in the box, she now thinks that box contains frightening possibilities. If this is a possibility, set up another box near the first and see if your cat will use it. If it seems likely the startling event will occur again, set up another box in a quieter environment as well. Then, implement the three-step program I described under "House-soiling" at the beginning of this chapter.

Inconvenient Feline Preferences

Cat Prefers Your Potted Plants

To prevent your cat from using potted plants for a litter box, and from playing in the dirt, simply keep your cat out of your potted plants. If you can, hang all of them from ceiling hooks. If that's not possible, cover the surface of the soil with decorative stones, mothballs (cats hate the smell), pine bark chips, wire mesh or aluminum foil. You can also cut strips of non-toxic, double-sided Sticky Paws for Plants (safe for cats and plants) and place them in crisscross patterns on the pots, or place them in the top of the pot at the base of the plant. Doing this won't interfere with watering.

Cats who are occasionally allowed to go outdoors to eliminate often try to use indoor houseplants for their toilet areas. You can try mixing a little potting soil, or soil from the garden, with the regular cat litter, to entice your cat to use her litter box. This works with some cats but encourages others to prefer the soil.

Urinating in the Bathtub, Sink or Shower

Urinating on hard, cool surfaces—the bathtub, sink, shower and tile floors, for instance—is one of the signs associated with feline lower urinary tract disease (FLUTD, see chapter 2). Consult your veterinarian immediately because FLUTD can be life-threatening. A cat who has been successfully treated for FLUTD may still prefer the sink or tub simply because that is what she has become accustomed to.

Once medical causes are either treated or ruled out, the solution is to limit the cat's access to these areas, while offering a very clean and attractive litter box as an alternative. Keep the bathroom door closed. If it must be left open, keep the shower door closed, fill the bathtub or sink with two to three inches of water for several days and place a clean litter box nearby. You can also place something the cat doesn't like to walk on, such as a sheet of aluminum foil or plastic bubble wrap, in the sink or around the counter. Bonnie Beaver, a behaviorist at Texas A&M Veterinary School, suggests placing an object such as a cactus plant in the sink for several days, with a clean litter box nearby. Ouch!

Stress-Related Problems

House-soiling and urine spraying can be indicators of stress that is affecting your cat. Anxiety and fear are common feline emotional reactions that cause many behavioral problems, including litter box problems. Stress may also trigger bouts of FLUTD. Even healthy cats can become anxious or frightened by changes that might seem insignificant to you. Some common indoor and outdoor stressors include:

- Changes inside and outside the home, such as adding new furniture or carpeting, redecorating or remodeling
- Introducing a new spouse or roommate into the home (see chapter 6 for advice on how to minimize stress when doing this)
- Introducing a new pet or baby into the family (see chapter 6 for advice on how to minimize stress through this transition)
- Moving to a new home
- Separation anxiety: vacation, hospital stay, returning to work or school and other situations that cause a cat to be separated from her family
- The loss of a companion cat or other pet
- Thunderstorms or continued exposure to loud noises
- Territorial conflicts and aggression in multi-cat households
- Visual contact with strange cats outside the home
- Geriatric-related problems

Changes Inside and Outside the Home

Cats like a secure and steady environment, and disruptions of any kind can be very stressful. They feel comfortable with familiar surroundings and the same old furniture. Making changes in home territory—even just moving the old furniture around—can provoke some cats to begin eliminating outside the litter box. Adding new chairs, a sofa or new carpeting changes the smell and texture of a cat's territory and can be intimidating because the items in question don't contain the cat's scent.

If you are planning to buy new furniture or carpeting, place a Feliway diffuser in the room to calm the cat and discourage her from marking or scratching. Once the new chair or couch is in place, take a towel or blanket that contains your cat's scent and put it on the

piece (or wipe it down with the towel) to accelerate the familiariza-
tion process. In addition to using the Feliway diffuser, you can cover
the piece with a sheet or blanket for a few days if you're worried that
the cat may be destructive.

If you're remodeling inside or outside your home, isolate the cat
as much as possible by confining her to a quiet room—one that has
not been remodeled and does not contain new furniture or carpets—
and place her litter box there. Make sure indoor workmen know
there is a cat present and that her room is off-limits. No one should
accidentally open the door and give her a chance to escape. Turn on
the radio or television (low volume) to let the cat hear the sounds
of music or human voices instead of construction noise. Pull down
the shades or close the curtains so she can't see outdoor workmen.
Keep her confined until the workmen leave. When they do, use the
same plan for a new piece of furniture to help your cat adjust to the
new room.

Recent Declawing

Declawing is the removal of the last joint of the toe, which
includes the bone attached to the cat's claw. It is performed
under general anesthesia and is an extremely controversial
procedure. One of the effects of the operation is moderate to
severe post-operative pain that will continue for several days
to weeks after surgery, making it difficult for a cat to scratch
and cover her waste in the litter box. During this period, and a
bit longer if necessary, veterinarians recommend replacing the
cat's regular litter with one of the lightweight pelletized litters
made from recycled paper or simply using shredded newspaper. This is because litter granules can irritate the paws. A few
days before the surgery takes place, mix some of the pelletized newspaper in with your cat's regular litter to ease the
transition. If you use shredded newspaper, it must be completely changed and the box rinsed and dried *each time* the
cat uses the litter box.

Moving to a New Home

Moving to a new home is very stressful for a cat. Felines are extreme-ly territorial, and your cat has probably identified your present house, or a part of the house, as her home territory. Moving to new sur-roundings can make a cat feel *very* insecure. Here are some sugges-tions to help make your cat more comfortable during the transition.

- If you don't own a cat carrier, buy one several weeks before you move. Leave the door open to allow the cat to go in and out at will, or to sleep inside. She should be familiar with the carrier before you move.
- While you are packing, confine your cat to a quiet room.
- On moving day, spray the carrier and your car with Feliway before you put the cat inside and transport her to the new home. Make sure the cat is safely in her carrier *before* the moving men arrive.
- While the movers are at the new residence, confine your cat to a quiet room with the door closed to keep her safe. Plugging in a Feliway diffuser in the confinement room at least an hour before the cat arrives will help calm her.
- Place her litter box, food and water dishes (not too close to the litter box) and her bed in the room, along with some favorite toys.
- Visit her often to reassure her that you're not abandoning her, but keep her confined until the movers have gone. Then let her gradually explore the new house.
- Protect your cat from any danger while you are unpacking (you may need to keep her out of the room), and check *all* boxes and wrapping paper for hiding felines before you dispose of them.
- Some cats feel totally overwhelmed in a new home, especial-ly if it is much larger than the old one. If your cat seems fear-ful and keeps close to the ground or is reluctant to explore her new surroundings, don't force things. She may feel better being confined to just one or two rooms for a few days. Make sure to stick to her routine as much as possible and spend a lot of time with her.

Separation Anxiety

Cats can suffer from anxiety when they are separated from their homes or from important people in their lives. If you are returning to work or to school after being home for most of the day, there's good news. Feline behaviorists say cats generally do not exhibit separation anxiety during an ordinary 8- to 10-hour period but will show signs if the owner is absent for 24 hours or longer. If you are going on vacation, try to have someone come in more than once a day and do more than simply feed your cats and scoop the box. Regular playtime and attention, even from a stranger, will help alleviate the stress your cat feels when you are gone.

Separation anxiety behavior may include urinating or defecating in inappropriate places, urine spraying, vomiting, not eating or drinking while the owner is away, vocalizing (crying and meowing), excessive self-grooming, scratching furniture and other destructive behavior. According to feline behaviorist Dr. Nicholas Dodman, "Certain Oriental breeds, such as Siamese and Burmese, may be more prone to develop separation anxiety than cats with more robust temperaments, like Maine Coons." Here are some steps you can take while you are away, even for a few hours:

- Play with your cat before you leave. Gradually wind down the play session, so the cat is tired and not keyed up when it's time for you to go.
- Leave the radio or television on (at low volume—cats have excellent hearing) to let the cat hear human voices.
- Leave a piece of unwashed clothing that you have worn in the place where the cat sleeps.
- Provide the cat with a window seat or tall cat tree that gives her an interesting view of the lawn or garden. Adding bird feeders near the window will make things more interesting.
- Prevent boredom by leaving an assortment of interesting toys, especially things with catnip, felt mice that can be hunted as prey and fuzzy things that are durably stitched with no small parts than can be bitten off and swallowed.

- Put half the day's ration of dry cat food in a toy that dispenses food as it is manipulated by the cat. Leave this "food puzzle" down only while you are away and pick it up as soon as you return.
- Play with your cat as soon as you return home.
- Use Feliway spray or diffuser to help reduce your cat's anxiety while you are away.

Loss of a Companion

A cat can be devastated by the loss of a beloved human or animal companion. Signs of feline mourning include loss of appetite, pacing (often accompanied by meowing and crying), lethargy, changes in sleep cycles or inability to sleep, disinterest in normal activities and toys and self-mutilation or sudden changes in self-grooming. Inappropriate elimination can also result.

This is the time to give your cat as much love and attention as you possibly can. If she doesn't feel like playing, just sit with her and stroke her in her favorite spot. Brush her more often if she is not grooming, and tempt her to eat with her favorite foods. Let her know you are there for her.

Extreme cases of anxiety due to the loss of a beloved companion may require the assistance of a veterinarian or a feline behaviorist.

Thunderstorms and Loud Noises

Some cats become stressed during thunderstorms or when they are exposed to prolonged loud noises. Unlike dogs, who can become extremely phobic at the first thunder clap and do themselves harm, cats usually react by hiding under beds, in closets or in remote corners, emerging unharmed when the storm is over. If possible, pick up the cat and confine her to a room where the lightning and thunder are barely visible and/or audible. Spraying Feliway or using a Feliway diffuser in the room can also help reduce her anxiety. During the storm, try to distract the cat by offering her favorite treats. Severe reactions are rare and may be treated with anti-anxiety medications under veterinary supervision.

Geriatric-Related Problems

House-soiling is a common problem of older cats, and many of the behavioral changes observed in older cats can be attributed to medical causes. FLUTD, kidney or liver disease, colitis, diabetes and other medical conditions that appear more frequently in older cats can cause pain during the process of urination and defecation. If your older cat starts having litter box lapses, have your veterinarian examine your cat to rule out any medical problems.

Stress can also cause house-soiling because older cats are generally not able to handle stressful situations as efficiently as younger cats. Gradually reducing many of the stress-related problems discussed in this chapter will greatly benefit both you and your cat. Using Feliway can also be helpful in reducing anxiety in older cats and subsequently decreasing house-soiling.

Older cats may need to use their litter box more often because of diminished bladder and bowel control. They may suffer from incontinence—an inability to control the sphincter muscles of the anus or urethra. This problem occurs more often in females, but it may affect cats of either sex. Some types of incontinence can be controlled with medication, so talk to your veterinarian. And remember that your old friend can no longer control herself.

Because of pain, weakness or arthritis, many older cats don't like to walk up and down stairs (especially if they are uncarpeted or steep) to use a litter box. Don't expect your older cat to go up and down stairs, or even to the other side of the house, to get to her litter box. Deviations in litter box habits often occur when older cats can't reach their boxes in time. Instead, place extra boxes in several rooms—especially in the rooms where your senior cat likes to rest. (But keep the box on the other side of the room from the cat's sleeping area because cats do not like to eliminate where they sleep or eat.) An older cat may also find it difficult to step into a litter box with high sides. In this case, provide a low-sided tray or create a small, sturdy ramp to make entry and exit easier.

Older cats sometimes become confused and forget where their litter boxes are located. If your cat does this, you may have to confine her to a room that is easy to clean when you are not around to

supervise. Elderly cats may totally forgo the litter box and start urinating in potted plants or other unsuitable places. Accidents may
happen in various parts of your home away from the litter box. *Don't
scold or punish an old cat for an accident that may now be beyond her
control.* Be understanding and make the litter boxes more accessible.

Some older cats become extra-fastidious about cleanliness; you
may need to change the litter in the box more often.

Do inspect the contents of the litter box often, and immediately report to your veterinarian any change in the frequency of urination or defecation; in the color, consistency or odor; or any loss of
continence.

Problems Between Cats

Veterinarians agree that there are fewer inappropriate elimination
problems in single-cat households. The more cats who live in a
home, the more likely one or more of them will house-soil or spray.

To avoid being repetitious, in addition to the advice offered
under each specific problem, the best solution in many of these cases
is neutering. Neutering decreases anxiety, territoriality and aggression and helps solve many behavioral problems.

The Territorial Imperative

As I explained in chapter 1, cats are territorial by nature. In the
wild, a cat's territorial boundaries are firmly established. When an
unknown cat is outside these boundaries, he will be ignored; inside
them, the stranger must be prepared to challenge the resident cat.
Indoor cats also establish their own territories within the house or
apartment where they live—a home range that includes favorite
places for sunbathing, dozing and sleeping and for sedately watching
the world go by, writes Bruce Fogle in *The Cat's Mind*. In most cases,
cats are completely familiar with all the nooks and crannies of their
immediate home territory. "Cats create a ranking hierarchy among
themselves," Fogle adds, "but it is neither as obvious nor as well
defined as it is with pet dogs. A closer examination, however, will
reveal that there is a pronounced social order."

Fortunately, except for minor spats, most cats adapt well to living indoors with other cats. This may be partly due to the fact that food is readily available since, according to a report in the Tufts University School of Medicine newsletter, studies show that cats' territories shrink as food becomes plentiful. It's no longer necessary, as is the custom in the wild, for a cat to defend his territory in order to have access to prey. As a result, most households can harmoniously support several cats.

Depending on the size of the home or apartment and the temperaments of the cats living there, however, eventually a point may be reached where the territory is no longer large enough. A confrontation indoors can become serious when one cat invades another's territory. And once aggression occurs, it tends to gradually get worse. When a territory is threatened, a dominant cat will usually try to defend his turf and intimidate the subordinate cat by aggressive behavior: swatting, hissing, growling and screaming vocal threats, baring the teeth and assuming certain offensive body postures (see chapter 1). If these postures and physical intimidation do not scare off the intruder, the dominant cat may resort to other destructive practices, including urine spraying or house-soiling, to reconfirm precisely just who is the king of the castle.

Male cats are generally more aggressive toward other cats, but females can also be quite contentious, especially when they have kittens or when another female is being introduced into the home. Take these steps to help reduce territorial conflicts:

- Always introduce a new kitten or adult cat to resident cats gradually (as discussed in chapter 6).
- Neuter all cats.
- A dominant cat will often try to keep a submissive one from using the litter box. Provide enough boxes—one per cat, plus one extra—and place them in separate locations.
- Provide adequate entry and escape routes around boxes to prevent a submissive cat from being trapped or ambushed.
- Have a separate feeding dish for each cat, and place them in separate locations.

- Using Feliway spray or room diffuser will help ease anxiety and the need to claim territory. Clinical studies show a 70 percent success rate for Feliway in reducing or eliminating spraying in multi-cat households.
- Severe situations require professional assistance. Consult a feline behaviorist about a behavior modification program. Some cats respond favorably to anti-anxiety drugs. Medication use should always be carefully supervised by your veterinarian.

Several Cats Sharing a Single Litter Box

Many cats don't mind sharing their litter box with other cats in the house, but some do. Some may be attracted to the scents left by their housemates, while other, more sensitive cats are repelled by the odors and may avoid using the box completely. There are also some cats who prefer to use two litter boxes for elimination—one for urinating and one for defecating. The general rule is to have one more litter box than the number of cats in your home. In other words, one cat should have two boxes, two cats should have three boxes, three cats four, and so on. Scoop all boxes several times a day and clean and refill them regularly.

One Cat Ambushes Another

A dominant cat may try to prevent other cats from approaching a particular litter box because it is located within his territory and the intruder is invading his space. In a multiple-cat household, space litter boxes far enough apart to avoid territorial confrontations. A cat, especially one who is nervous or timid, can also be ambushed while he's in the litter box or exiting it by another cat, a dog or a child. This may make him nervous enough to seek another place to eliminate. If this is happening, make sure there are several ways to get out of the box. This means removing any cover and locating the box so it is not backed into the corner of a room. This way, if the bully is lurking in one area, the timid cat is not trapped and can escape via another route.

Visual Contact With Other Cats Outside the Home

The sight or smell of another cat, particularly a tom or a female in heat—observed through a window or a screen door or in the hall-way of an apartment building—can trigger an outburst of territorial urine spraying. To solve this problem:

- If you have cat trees and perches, or chairs and sofas, placed near the windows, temporarily move them away.
- Close the curtains, pull down the blinds or block the view from a screen door so the cat can't see or interact with the outsiders.
- Remove things in the yard—bird feeders, food, garbage cans—that might attract stray cats.
- Turn on the sprinkler system or spray a repellent—*non-toxic to cats*—around your home to deter roaming cats from coming onto your property.
- Clean any urine marks made by the intruders on the outside of your home with an enzymatic cleaner that will remove all traces of the odor.
- Spray Feliway one to two times a day on prominent areas in your home near windows or doors (cat window seats, window frames, door frames) and urine marks on walls, to reduce your cat's stress and the need to claim territory. Before spraying the walls, clean the urine marks with clear water only, as detergents and other cleansers may degrade the product.

When a Cat Rejects the Litter Box

Bear in mind that urination and defecation are significant ways animals communicate. If your cat starts eliminating outside her litter box, she is doing so for a reason—to register her displeasure. Something has changed abruptly and dramatically in her life. The cause may be medical, physiological or behavioral. Once the cause is determined, most litter box problems can be corrected by either medical treatment or some behavior modification. *Remember, punishment is not the answer.*

I have only scratched the surface in explaining the reasons cats avoid eliminating in their litter boxes. If you wish to do more research about specific problems online, I highly recommend www.petplace.com, a site supported by Angell Memorial Animal Hospital. There, you can access the latest information on behavior, nutrition, general cat care and illness, written by experts but in laypersons' terms.

Do remember that the longer you ignore house-soiling and urine spraying problems, the more serious they will become and the more difficult they will be to solve. Keep in mind, too, that what works for one cat may not be the best solution for another. If problems persist, talk to your veterinarian or seek the advice of a feline behaviorist. But please, don't give up on your cat.

6

Proper Feline Introductions

Cats are creatures of habit, and any change in their routine can cause stress. And, as I mentioned in chapter 5, stress can trigger inappropriate elimination. Spreading familiar smells around the house is one way cats comfort themselves. They also use their urine and feces to reinforce the boundaries of their territory when they feel is it under threat. Some veterinarians now believe stress can trigger bouts of FLUTD (see chapter 2), as well.

Proper, gradual introductions can go a long way toward relieving the stress of changes in the household. In this chapter I'll explain how to bring new people and new cats into your home in ways that cause the least amount of upset for your cat. You should also try to stick to your cat's normal routine as much as possible. You'll find other suggestions for reducing feline stress in chapter 5.

Introducing a New Spouse or a Roommate

Whether they're a spouse or a roommate, a new person in the home is a stressor for cats. Their anxiety often causes them to urinate or defecate on top of specific items, such as your bed, a favorite chair or articles of clothing or even in shoes. A cat who does this is not being vindictive—he is simply telling the newcomer, "These things are in *my* territory."

One of the best solutions is to have your new spouse or roommate interact with the cat *before* moving in. Don't pressure either one; let them to get to know each other and bond on their own terms. During this period, ask the new person to leave an unwashed

T-shirt or other article of clothing at your home, so the cat can get used to their scent. Spray the new person's shoes and clothing with Feliway. To encourage a positive relationship, ask the new person to feed the cat (and remain in the room at a distance while the cat eats) and to play with the cat regularly.

Does the cat sleep with you in bed? Is the cat allowed on the dining room table or the kitchen counter? Do you feed the cat tidbits from the table? Your new spouse or roommate may object, or may be allergic to cats, and compromises may be necessary.

If you must keep the cat out of the bedroom at night, have a fun, interactive play session with your cat just before bed. Wind down the play slowly and then go into your bedroom and close the door. You may be a bit stressed when the cat meows outside the bedroom door, but find him another warm and comfortable place to sleep and shower him with affection, and matters should eventually work out.

Don't be discouraged; it may take several months for everyone to feel comfortable. And be sure to thoroughly launder all clothing and bedding stained by cat urine or feces, and use an enzymatic cleaner to remove stains and odors on carpeting and furniture.

Bringing Home a New Baby

Once again, the key to introducing a newcomer into a household is to give your cat a chance to get used to the change gradually and to make sure he doesn't feel left out. A new baby who is suddenly receiving lots of attention can make the resident cat stop using his litter box, especially if he has been the only pet or "baby" in the family before the new arrival. As wonderful, exciting and overwhelming as a new baby is, if you ignore your cat at this time, he's bound to resent it, become stressed and associate your sudden aloofness with the baby.

But while you can have a new spouse or roommate meet the cat several times before they move in, that's not possible with a baby. So how do you make a gradual introduction? It just takes a little advance preparation, patience and common sense.

Start by bringing home all the furniture and baby products well in advance of the baby's arrival. Let the cat thoroughly check out the baby's room. Open packages of diapers, baby lotion, wipes and all the other things you will use, place them on the floor of the baby's room and let your cat become familiar with the strange smells. Dab a little baby lotion or powder on yourself, so the cat comes to associate those smells with someone he already knows and loves.

Next, desensitize your cat to the sound of a baby's crying. Dr. Peter Kross, a Manhattan veterinarian, suggests asking friends who have babies to tape their cries, and play the recording frequently at home at a gradually increasing volume. If possible, have your friends bring their children to visit your cat, supervising them closely so they don't catch hold of him or pull his tail.

After the birth but before you bring the baby home from the hospital, bring home a blanket or a piece of clothing with the baby's scent on it and put it in a place the cat will be sure to sniff.

Soon after you bring the infant home, sit in a comfortable chair—preferably in the baby's room—and hold the baby in your arms. Encourage your cat to come closer. Let him sniff the baby and thoroughly explore the room so that he can become accustomed to the infant's scent. *Do not leave your cat alone with the baby.* However, as you attend to the baby, do let the cat be in the room, if possible, and pay lots of attention to him. Talk to the cat as you change or dress the baby; tell him that you love him as much as ever. This may sound silly, but he'll understand your intent by the tone of your voice.

Always supervise your cat's interactions with your baby. If you are concerned about contact, you can buy a tentlike "cat net" at a baby supply store and place it over the crib to prevent the cat from getting inside. If your cat shows *any* hostility to the baby, you *must* keep them apart. The chances of a cat suddenly attacking a baby are very rare, but if your cat becomes aggressive, he should be firmly and immediately taken out of the baby's reach. Discuss this problem with your veterinarian, as anti-anxiety medications may help to calm him.

Bringing Home a New Cat

Introducing a new kitten or adult cat into a household is usually a traumatic experience for the other pets in residence—and emotional upset is a major cause of house-training lapses. Gradual, carefully planned introductions *always* go better than simply bringing home a new cat and waiting to see what happens.

Cats can and do establish successful relationships with other cats and dogs, especially when they are not just forced together helter-skelter without proper introduction. You must be mindful of the nature of the pets you are putting together. Knowing the resident pet and his temperament—active, docile, dominant or aggressive, for instance—will play an important role in a smooth transition.

It's generally less troublesome to introduce a kitten to a resident adult cat or dog. Adult cats will adapt more easily if they were exposed to other cats or dogs when they were young. Unfortunately, cats who were weaned too early and raised with no contact with other cats have more problems getting along with other pets. In any case, you should always consider the emotions of the resident pet and make every effort to keep him from feeling that you're favoring the newcomer.

Always keep the new pet and the pets in residence separated at first. And never leave your new kitten or adult cat unsupervised, especially around a dog, or he could be injured.

Do anticipate a little minor squabbling, hissing, spitting, swatting and other posturing when you add a new kitten or cat to your home, no matter how docile and well-adjusted your resident pets are. It's normal. Just in case a fight does break out, keep a spray bottle filled with water in readiness. As soon as you see any aggression, immediately squirt the aggressor (not in the face, please!). This startling action should be enough to break up the fight. Eventually, just reaching for the spray bottle will produce the desired results. Matters will also proceed more smoothly if the pets have been neutered (the females spayed and the males castrated) as these procedures do make both species and sexes less aggressive.

Preliminary Steps

Confining a new cat was described in detail in chapter 4. Many animal behaviorists suggest that during the new cat's confinement, a good precursor to fact-to-face contact with resident pets is to feed the newcomer and the veteran on either side of the confinement room door. This way, they can smell and even hear each other while they are doing something they love—eating.

It's also a good idea to give the resident cat or dog a towel or a small blanket that the new cat has slept on, and to give the new cat something your resident pet has used, so that they can become familiar with each other's scent. As soon as the new kitten or cat is appropriately using his litter box, you can start letting him out of confinement to gradually explore the rest of the house. Sometimes simply put the resident pet in the new cat's room, and the new cat in a room the resident pet spends a lot of time in. If you have a dog, let the new cat wander around your home while the dog is out with another family member. When he does so, his scent will start permeating the house and begin to become familiar. Each time the new cat has the run of the house, however, you must make sure the resident pet is away or confined. Give the resident pet lots of praise, love and attention so that there's no jealousy or depression—which can also result in inappropriate elimination behavior.

Now the introduction process can begin.

Cat Meets Cat

Cats are territorial creatures, and they can be quite resentful when a new kitten or adult cat moves into their home. The resident cat is not being asocial—he's defending his territory. Therefore, when you have other cats in residence, a new kitten or cat must be introduced gradually. And even when the introduction process proceeds slowly, don't expect an instant peaceful relationship. Some cats quickly learn to live together harmoniously, while others endure one another contemptuously for months and then, eventually, become friends.

Give each cat individual attention and stroking sessions while the introduction process is taking place. A resident cat who has been, up till now, the only pet in your household will be depressed when he doesn't receive the same or even more attention than the new cat. He may react by behaving inappropriately—urinating or defecating outside the litter box, for instance. So make sure he knows he still has a special place in your heart.

Here are some more tips for smooth introductions

- Let the cats become familiar with one another by scent first, as described in the previous section.
- After a few days of confinement, put the new cat in his carrier, open the door to the confinement room and bring in the resident cat to meet the newcomer. Don't open the carrier. Allow the cats to smell and investigate each other through the carrier without a physical confrontation. Make a huge fuss over both cats. You may hear some hissing or growling at first, but they probably will only be checking each other out. If there is any aggressive posturing (spitting, tail swishing, fur bristling) don't be alarmed—the new cat is safe in his carrier. The more often the pets casually meet, the more they will become accustomed to each other, and the fussing will eventually stop.
- The next time the cats are introduced, put the resident cat in the carrier and allow the kitten the freedom of the room. Keep alternating meetings in this manner for a few days.
- When they seem calm, let the cats meet and approach each other outside the carrier for short periods of time. Don't enclose the cats in one room when they meet; give them space to escape to another part of the home if they wish, to avoid a fight. Keep a spray bottle filled with water close by. If a fight breaks out, squirt the aggressor as previously described.
- As soon as the cats have accepted each other, feed them at the same time but in different areas (it can be in the same room but at opposite ends), and place the new cat's litter box in a different location than the resident cat's. This helps avoid competition. The cats will also need separate toys and individual attention from you.

Cat Meets Dog

People often assume cats and dogs are born to fight, but in reality they can live together harmoniously. As I mentioned in chapter 1, dogs are social animals who naturally live in packs, and they can accept other species into their pack. Pack relationships are based on a hierarchy in which each member of the group occupies a certain rank or position. There is always a leader—called the *alpha*—who controls the pack and keeps everyone in line, a dog number two, dog number three, and so on down the line. The same pack behavior patterns govern a dog's relationships with other species, so cats become members of the pack. In fact, in most households with a dog and a cat, the cat often ends up higher in the pack hierarchy than the dog.

One advantage of owning a cat and a dog is that their territorial imperatives generally don't conflict the way interactions between two or more cats can. Matters will also proceed more smoothly if your dog is obedience trained and understands the basic commands *sit, stay, down, come* and *no.*

- Let the two pets become familiar with each other by scent first, as described previously.
- Trim both the dog's and cat's nails before any face-to-face meeting. Remember that a cat's claws are an important part of his natural defenses, so you don't want even a young kitten scratching the dog in an attempt to fend her off.
- Immediately before the meeting takes place, take the dog outside. Play with her and give her plenty of exercise to tire her out a little and make her less exuberant.
- Feed both the dog and the cat before the meeting. Both will be less lively on a full stomach.
- Once back indoors, attach the dog's leash to her collar so you have complete control of her. Have a friend or a family member assist, if necessary.
- Bring the cat, in his carrier, and the dog together in a room. Put the carrier down and let each animal look at and sniff each other. Talk quietly and reassuringly to both pets. Repeat this step over several days until both animals are tolerating each other without any hissing or growling.

- Now it's time for a little closer contact. Command the dog to *sit* and *stay*. Treats can be offered when he obeys the commands. Bring the cat (in his carrier) into the room and open the carrier door. Let the cat come out on his own and get close enough to the dog to let them investigate each other. Don't force the cat to come out and meet the dog. If he does, keep the dog on a leash and let them sniff each other while you remain calm.

- Repeat this step over several days, gradually extending the frequency and duration of the meetings until the cat and dog are comfortable in each other's presence. *Until both pets completely accept each other, always supervise face-to-face meetings.*

- Do anticipate that something may go amiss: make sure there is a cat tree, the top of a table or chest or another high surface in the meeting room for the cat to safely get out of the dog's way if he needs to. Correct any attempts by the dog to chase the cat. Use the water spray bottle if there is any aggression.

- *Always* keep both pets apart when you go away from home until you are certain the dog has accepted the cat—and vice versa.

7

Is That *Eau de Cat* I Smell?

Is there anything less appealing than walking into a home and smelling the disgusting stench of cat urine? Yuck! It's nauseating—and it smells even worse when the weather is wet or humid. When a cat stops using her litter box for any reason, she looks to find another surface in the house that feels right to urinate or defecate on.

Odors and stains caused by inappropriate elimination and urine spray marking can be major problems for cat owners because they often go undetected for long periods of time while they build up. Meanwhile, the cat keeps returning to re-soil the same location, and the hidden stain keeps growing and growing. When the problem is ignored, urine can seep through the carpeting, into the padding and eventually to the floor underneath and can become difficult if not impossible to remove. Worse yet, urine deposited outside the litter box may inspire other cats in the household to be less discriminating about where they eliminate.

What Causes the Odor?

Cat urine is said to be the worst source of pet stains and odors. Male cat urine smells the strongest, but females excrete pungent urine too. Another consideration is the cat's age: the older the animal, the stronger the odor.

Odor problems are generally caused by ammonia, mercaptans and other chemicals and gasses emitted from urine and feces. Fresh urine has very little odor, but as soon as it is deposited onto a surface, it begins to decay. In the first stage of decay, the urea in the urine breaks down into amines (ammonia) that create an unpleasant odor. The second stage of decay produces mercaptans that cause the most serious odor problems. Mercaptans are the chemicals that give skunk spray and rotten cabbage their distinctive odors. If they aren't eliminated during the cleaning process, the urine odor will remain and become stronger over time. That's why it is most important to clean the carpet, furniture, floor or any other spot on which your cat has chosen to relieve herself as soon as possible. Left untreated, odors will smell worse and stains will get bigger.

The cat's sense of smell is highly developed—many times greater than ours—and if even *one* drop of urine remains on a carpet or floor, your cat will detect it because her nose is close to the floor. Timing is vital! The sooner you eradicate the stain and odor, the easier it will be to keep your cat from returning again and again to the area in question. And the longer a stain sits, the more difficult it becomes to remove.

There are many effective pet stain and odor removers on the market that, when used according to directions, instantly destroy the odor so even your cat cannot smell it. Most contain chemical products, enzymes or bacteria, alone or in combination. Chemical products modify the urine and lift the stain. Enzymatic and bacterial products metabolize the organic compounds in urine and literally digest the bacteria that cause odors. They break down the odor and the stain and remove both completely. These products are non-toxic, biodegradable and safe for use around pets and children.

You have several choices when it comes to products. Nature's Miracle Stain & Odor Remover, Urine-Off Odor & Stain Remover, Lambert Kay Fresh 'n Clean Odor & Stain Eliminator with Oxy-Strength, Drs. Foster & Smith Stainaway, Outright Pet Odor Eliminator, Urine Erase, Simple Solution Stain & Odor Remover and Anti-Icky Poo are a few extremely effective brands that you can find in most pet supply stores or from the mail order catalogs listed at the end of this book. You can also ask your veterinarian or pet supplies dealer for their recommendations. Because you should clean the stain immediately, it's best to have a pet stain cleaner on hand, just in case.

Cleaning the carpet with water, club soda, vinegar, household cleaners, Scope mouthwash or ordinary carpet cleaners will only cover up the smell for a few days. *None* of these is as efficient as enzymatic or bacterial products specifically formulated to eliminate stains and odors from pet accidents. In any case, you should never use ammonia-based products or vinegar to clean pet stains because they leave a residue that smells like urine—which will only encourage your cat to return to the same spot to eliminate again and again.

Cleaning Hard Surfaces

Tile, slate, sealed concrete, seamless vinyl, linoleum, laminated flooring, semi-gloss or gloss enamel paint and other surfaces that don't absorb liquid are easy to clean, advises Don Aslett, a well-known cleaning expert. Simply soak up the urine with paper towels, mop the floor with a deodorizing cleaner made to remove pet stains and odors, rinse and let dry.

Porous surfaces, such as unsealed concrete, unfinished wood and flat or matte-finish latex paint, will absorb some urine. Urine can also seep into the seams between vinyl tiles and cause the odor to remain. In this case, wipe up all the excess urine, mop the floor with a detergent cleaner, rinse and follow with an odor neutralizer. If a strong odor still remains, you may need to reseal or repaint the surface after deodorizing to prevent your cat from re-soiling that spot.

Cleaning Carpets

Follow these steps to treat wet stains on carpets (these instructions will also work on most upholstery fabrics):

1. Remove all the urine from the area. *Blot*—do not scrub—damp areas with white paper towels or old terry-cloth towels to soak up as much of the liquid as possible. Always blot the carpet because rubbing or scrubbing can damage the carpet fibers, weaken the material and spread the stain. Keep repeating this step until the towels are completely dry when you

press on the spot. If necessary, press down with your heel or place something heavy, such as a few books or a brick, on the towels to help absorb the liquid.

2. Dilute the stained area with distilled water. After a thorough rinse, remove as much water as possible by blotting again with paper or terry-cloth towels or a wet vacuum. Because you will follow this step with an enzymatic or bacterial cleaner, it's best to use distilled water as enzymes can mix with the minerals in regular tap water and discolor the carpet.

3. After you have diluted and blotted up as much urine as possible, the next step is to neutralize the odors by spraying or pouring on an enzymatic or bacterial pet stain and odor neutralizer. Be sure to test a hidden area of the carpet to be treated for colorfastness beforehand. Enzymatic and bacterial products are not just perfumed cover-ups; they are designed specifically to penetrate deep into the carpet fibers and the padding underneath and to "eat" or absorb the odors and remove stains.

4. Soak the soiled area of the carpet thoroughly with the cleaner. Wearing rubber gloves, work the solution in with your fingers through the carpet and into the padding and floor below. The point is to use enough of the neutralizer to get into the carpet as deep as the urine did. Generally, when urine reaches the carpet padding, it spreads—so much so that what may look like a small stain on the surface may be twice that size down in the padding. Carpet experts generally recommend treating an area two to three times larger than the visible stain. Otherwise, the odor will remain and tempt your cat to return to the same spot.

5. No matter what stain and odor remover you decide to use, always follow the manufacturer's instructions carefully regarding how long the solution should remain on the carpet before you blot it dry. The carpet should be dried thoroughly to prevent the padding from rotting.

6. If you notice the stain reappearing on the surface of the carpet within a week or so, a second treatment may be necessary.

Wood Floor Stains and Odors

Cat urine is very damaging to wood floor finishes. When ignored, urine will seep through the carpet and padding onto the wooden floor underneath and can eat away even the toughest of wood finishes. You may need to lift the carpet and pad and then sand the surface and apply a new finish. The same is true if urine from spray marking has dripped down the wall into the baseboard and discolored the paint or varnish. Seek the advice of the professionals at your local hardware store, Home Depot or other building supply store regarding the supplies and methods necessary to bring your floors and walls back to their original condition.

Please remember: consult your veterinarian if you have to repeatedly clean up urine spots. Frequent house-soiling can be a sign of urinary tract infection, a serious condition that needs immediate treatment.

Old Carpet Stains

Dried, set-in stains on the carpet are a more serious matter. Apart from odor developing, the older the urine stain, the greater the chances of it having soaked through the carpet into the padding or wood floor underneath. According to the Association of Specialists in Cleaning and Restoration, cat urine can affect dyes used in carpeting. Not all deposits cause permanent stains, but the carpet dyes and finish, the content of the urine and the time elapsed after the urine is deposited are important. Urine can sometimes change the color of the dyes immediately, while other times it can take weeks or months for a reaction to become noticeable.

The likelihood and intensity of staining may be influenced by the cat's age, health, sex and reproductive cycles, medications and nutrition. Feline diets are higher in protein than canine diets, so cat urine produces stronger odors and is more likely to cause stains. Depending on these factors, some urine stains may be difficult if not

impossible to remove. You might need to remove or replace the carpet and padding.

   The best choice for treating an old, dried urine stain is a bacteria and enzyme neutralizer, as previously mentioned. "A bacteria/
enzyme digester is really the only way to completely eliminate the
organic material down deep in the carpet that causes the odor from
a pet stain in carpeting," Aslett advises, "especially one that's had a
chance to penetrate."

Can't Find the Stain? Try a Black Light

Nothing is more disconcerting than smelling the odor of cat
urine and not being able to find the source of the smell. Cats
are quite different from dogs when they have "accidents"
indoors. Dogs will lift their leg and urinate on your furniture or
mess on the middle of the carpet right in front of you, while
cats hide away their misdeeds behind the couch, under the
chair, under the bed or some other secret place that's difficult
to find.

   If your nose or eyes can't find the soiled area, a black light
can help you find it, even when urine accidents have gone
undetected for a long time. When seen under a black light, cat
urine that has been deposited or sprayed onto rugs and upholstery, floors and walls shows up as fluorescent, yellowish green
spots or drips.

   Black lights are rather weak, so to make it easier to identify soiled areas, turn off all lights in the room and close the curtains or blinds before you use one. Then, if necessary, outline
the areas that need to be cleaned with chalk. Incidentally, if
your carpet surface has been previously cleaned, it will be
harder to detect urine stains with a black light.

   Black light urine detectors can be purchased at most pet
supply stores, by mail or online from the pet suppliers listed at
the end of this book. Black lights are also available at Home
Depot, Wal-Mart and many electronics stores.

Special formulas that contain enzymes designed for deep penetration are available. Before applying the product, dilute the area with distilled water and blot thoroughly to help reduce the concentration of urine. You may need to repeat these steps on hard-to-remove stains, but if the concentration of urine is extensive and has set in for a long time, it may be virtually impossible to eliminate every trace of stain and odor. If you need additional technical skill, consider calling in a professional carpet cleaner for an opinion.

Removing Feces

When your cat has a bowel movement on the floor or any other hard surface, cleaning it up is relatively simple. If the stool is dry and formed, simply pick it up with paper towels. If the stool is semiliquid or completely liquid, blot the stained area with paper towels, mop with a deodorizing cleaner, rinse and let dry.

Removing feces from carpeting requires more effort. If the stool is solid, *carefully* lift it off the carpet using paper towels or two pieces of rigid cardboard or by slipping a small plastic food bag over your hand and picking it up. Lift the stool lightly; you don't want to push any part of it deeper into the rug. Once the stool is removed, treat the carpet with a bacterial or enzymatic cleaner, following the manufacturer's instructions carefully.

Semi-liquid or liquid (diarrhea) stools, which can be caused by anxiety, stress, excitement, intestinal parasites or dietary changes, are a more serious matter. When the stool is soft or fluid, odds are that some of the liquid has seeped down into the carpet fibers. The first step is to remove as much of the surface mess as possible by blotting the area with paper towels or by using an absorbent compound containing granules, such as Nature's Miracle Pet Mess Easy Clean-up Granules. Sprinkle the granules onto the soiled area, wait a few minutes until the diarrhea stiffens, then scrape it up with a metal spatula or push it into a dustpan. Pour on and saturate the area with a bacterial or enzyme cleaner, carefully following the manufacturer's directions. Repeat if necessary.

Another solution to the nasty problem of picking up liquid or semi-liquid waste is Poop Freeze, a non-flammable freeze spray. Simply spray and then wait 10 seconds while a frosty film hardens the waste and makes it easier to remove.

*A final reminder: If you discover an accident outside the litter box after it has happened, **never** rub your cat's nose in urine or feces, never yell at your cat and never strike her with your hand, a rolled-up newspaper or anything else. Cats do not associate past misdeeds with current punishment.*

Appendix

Recommended Resources

Books

Adams, Janine. *How to Say It to Your Cat.* New York: Prentice-Hall, 2003.

Aslett, Don. *Pet Clean-Up Made Easy.* Pocatello, Idaho: Marsh Creek Press, 1988.

Beaver, Bonnie V. *Feline Behavior: A Guide for Veterinarians.* Philadelphia: W. B. Saunders Company, 1992.

Dibra, Bash, with Elizabeth Randolph. *Cat Speak.* New York: New American Library, 2001.

Duno, Steve. *The Everything Cat Book.* Avon, MA: Adams Media Corporation, 1997.

———.*KISS Guide to Cat Care.* New York: DK Publishing, Inc., 2001.

Fogle, Bruce. *The Cat's Mind.* New York: Howell Book House, 1995.

Johnson-Bennett, Pam. *Think Like a Cat.* New York: Penguin Books, 2000. (If you're looking for solid information about all phases of rearing a well-adjusted cat, *this book is it!*)

Siegal, Mordecai (ed.). *The Cornell Book of Cats.* New York: Villard, 1997.

Turner, Dennis C., and Patrick Bateson (eds.). *The Domestic Cat: The Biology of Its Behavior.* New York: Cambridge University Press, 1988.

Periodicals

Cat Fancy (monthly)
Cats USA (annual)
Kittens USA (annual)
Fancy Publications
Subscription Service Department
P.O. Box 52864
Boulder, CO 80322-2864
(all three magazines are available at most newsstands or by subscription)
(900) 365-4421
www.animalnetwork.com

Catnip
Tufts University School of Veterinary Medicine
Subscription Services
P.O. Box 420235
Palm Coast, FL 32142-0235
(800) 829-0926
catnip@palmcoastd.com

CatWatch
Cornell University College of Veterinary Medicine
Subscription Department
P.O. Box 420235
Palm Coast, FL 32142-0235
(800) 829-8893
www.catwatchnewsletter.com

Web Sites

The Internet can be a valuable source of information about litter boxes and
their accessories, litter and inappropriate elimination. It can also be a great tool
to learn more about feline behavior, illness and wellness in general. Some excellent sites to consult are:

> www.cats.about.com
>
> www.catsinternational.org
>
> www.cleanreport.com
> (click on "Cleaning FAQ" for cleaning advice from Don Aslett)
>
> www.feline-behavior.com
>
> www.hsus.org
> (Humane Society of the United States)
>
> www.peteducation.com
>
> www.petplace.com
> (supported by Massachusetts SPCA Angell Memorial Hospital;
> first-rate, up-to-the-minute information)
>
> www.petsmarts.com
>
> www.thedailycat.com
>
> www.vetcentric.com
>
> www.vetinfo.com
>
> www.vetmedcenter.com

Where to Buy Equipment and Supplies for Your Cat

Most pet supply stores, especially Petsmart (www.petsmart.com) and Petco (www.petco.com), plus megastores like Wal-Mart and Target, stock a complete selection of litter boxes and accessories, litter, scratching posts, products for stain and odor control, kitty furniture, cat carriers, feeders, nutritional supplements, hairball remedies, training aids, Feliway, medications, toys and other items that will keep your cat's body and mind healthy, happy and active. You can also browse online and/or buy from the following mail-order pet supply catalogs. The companies listed below will send you an illustrated catalog on request. All accept major credit cards.

Drs. Foster & Smith
Just for Cats
2253 Air Park Road, P.O. Box 100
Rhinelander, WI 54501-0100
(800) 826-7206
www.DrsFosterSmith.com

PetEdge
P.O. Box 128
Topsfield, MA 01983-0228
(800) 738-3343
www.petedge.com

Care-A-Lot Pet Supply Warehouse
1617 Diamond Springs Road
Virginia Beach, VA 23455
(800) 343-7680
www.carealotpets.com

INDEX